Inverse and Ill-Posed Problems Series 54

Managing Editor
Sergey I. Kabanikhin, Novosibirsk, Russia/Almaty, Kazakhstan

Anatoly B. Bakushinsky
Mihail Yu. Kokurin
Alexandra Smirnova

Iterative Methods for Ill-Posed Problems

An Introduction

De Gruyter

Mathematics Subject Classification 2010: Primary: 47A52; Secondary: 65J20.

ISBN 978-3-11-025064-0
e-ISBN 978-3-11-025065-7
ISSN 1381-4524

Library of Congress Cataloging-in-Publication Data

> Bakushinskii, A. B. (Anatolii Borisovich).
>   [Iterativnye metody resheniia nekorrektnykh zadach. English]
>   Iterative methods for ill-posed problems : an introduction / by
> Anatoly Bakushinsky, Mikhail Kokurin, Alexandra Smirnova.
>   p. cm. − (Inverse and ill-posed problems series ; 54)
>   Includes bibliographical references and index.
>   ISBN 978-3-11-025064-0 (alk. paper)
>   1. Differential equations, Partial − Improperly posed problems.
> 2. Iterative methods (Mathematics)   I. Kokurin, M. IU. (Mikhail
> IUr'evich)   II. Smirnova, A. B. (Aleksandra Borisovna)   III. Title.
>   QA377.B25513 2011
>   515'.353−dc22
>   2010038154

Bibliographic information published by the Deutsche Nationalbibliothek

The Deutsche Nationalbibliothek lists this publication in the Deutsche Nationalbibliografie; detailed bibliographic data are available in the Internet at http://dnb.d-nb.de.

© 2011 Walter de Gruyter GmbH & Co. KG, Berlin/New York

Typesetting: Da-TeX Gerd Blumenstein, Leipzig, www.da-tex.de
Printing and binding: Hubert & Co. GmbH & Co. KG, Göttingen
♾ Printed on acid-free paper
Printed in Germany
www.degruyter.com

Preface

A variety of processes in science and engineering is commonly modeled by algebraic, differential, integral and other equations. In a more difficult case, it can be systems of equations combined with the associated initial and boundary conditions. Frequently, the study of applied optimization problems is also reduced to solving the corresponding equations. Typical examples include Euler's equation in calculus of variations and boundary value problems for Pontrjagin's maximal principle in control theory. All such equations, encountered both in theoretical and applied areas, may naturally be classified as *operator equations*. These equations connect the unknown parameters of the model with some given quantities describing the model. The above quantities, which can be either measured or calculated at the preliminary stage, form the so-called input data. Generally, the input data as well as the unknown parameters are the elements of certain metric spaces, in particular, Banach or Hilbert spaces, with the operator of the model acting from the solution space to the data space. The current textbook will focus on iterative methods for operator equations in Hilbert spaces.

Iterative methods in their simplest form are first introduced in an undergraduate numerical analysis course, among which Newton's method for approximating a root of a differentiable function in one variable is probably the best known. This is a typical iterative process widely used in applications. It can be generalized to the case of finite systems of nonlinear equations with a finite number of unknowns, and also to the case of operator equations in infinite dimensional spaces. It should, however, be noted that direct generalization of this kind is only possible for *regular* operator equations and systems of equations. The regularity condition generalizes the requirement on the derivative to be different from zero in a neighborhood of the root. This requirement is used for the convergence analysis of Newton's scheme in a one-dimensional case. Without the regularity condition, Newton's iterations are not necessarily well-defined. The lack of regularity is a major obstacle when it comes to applicability of not only the Newton method, but all *classical* iterative methods, gradient-type methods for example, although often these methods are formally executable for irregular problems as well. Still, a lot of important mathematical models give rise to either irregular operator equations or to operator equations whose regularity is extremely difficult to investigate, for instance numerous nonlinear inverse problems in PDEs. Thus, the question is whether or not it is possible to construct iterative methods for nonlinear operator equations without the regularity condition.

In the last few years the authors have been developing a unified approach to the construction of such methods for irregular equations. The approach under development is closely related to modern theory of *ill-posed problems*. The goal of our textbook is to give a brief account of this approach. There are 16 chapters (lectures) in the manuscript, which is based on the lecture notes prepared by the authors for graduate students at Moscow Institute of Physics and Technology and Mari State University, Russia, and Georgia State University, USA. A set of exercises appears at the end of each chapter. These range from routine tests of comprehension to more challenging problems helping to get a working understanding of the material. The book does not require any prior knowledge of classical iterative methods for nonlinear operator equations. The first three chapters investigate the basic iterative methods, the Newton, the Gauss–Newton and the gradient ones, in great detail. They also give an overview of some relevant functional analysis and infinite dimensional optimization theory. Further chapters gradually take the reader to the area of iterative methods for irregular operator equations. The last three chapters contain a number of realistic nonlinear test problems reduced to finite systems of nonlinear equations with a finite number of unknowns, integral equations of the first kind, and parameter identification problems in PDEs. The test problems are specially selected in order to emphasize numerical implementation of various iteratively regularized procedures addressed in this book, and to enable the reader to conduct his/her own computational experiments.

As it follows from the title, this textbook is meant to illuminate only the primary approaches to the construction and investigation of iterative methods for solving ill-posed operator equations. These methods are being constantly perfected and augmented with new algorithms. Applied inverse problems are the main sources of this development: to solve them, the successful implementation of well-known theoretical procedures is often impossible without a deep analysis of the nature of a problem and a successful resolution of the difficulties related to the choice of control parameters, which sometimes necessitates modification of the original iterative schemes. At times, by analyzing the structure of particular applied problems, researchers develop new procedures (iterative algorithms, for instance), aimed at these problems exclusively. The new 'problem-oriented' procedures may turn out to be more effective than those designed for general operator equations. Examples of such procedures include, but are not limited to, the method of quasi-reversibility (Lattes and Lions, 1967) for solving unstable initial value problems (IVPs) for the diffusion equation with reversed time, iteratively regularized schemes for solving unstable boundary value problems (BVPs), which reduce the original BVP to a sequence of auxiliary BVPs for the same differential equation with 'regularized' boundary conditions (Kozlov and Mazya, 1990), and various procedures for solving inverse scattering problems. For applied problems of shape design and shape recovery, the level set method is widely used (Osher and Sethian, 1988). The reader may consult [59, 27, 63, 69, 40] for a detailed theoretical and numerical analysis of these algorithms.

The formulas within the text are doubly numbered, with the first number being the number of the chapter and the second number being the number of the formula within the chapter. The problems are doubly numbered as well. A few references are given to the extensive bibliography at the end of the book; they are indicated by initials in square brackets. Standard notations are used throughout the book; $\mathbb{R}$ is the set of real numbers, $\mathbb{N}$ is the set of natural numbers. All other notations are introduced as they appear.

The authors hope that the textbook will be useful to graduate students pursuing their degrees in computational and applied mathematics, as well as to researchers and engineers who may encounter numerical methods for nonlinear models in their work.

Anatoly Bakushinsky
Mikhail Kokurin
Alexandra Smirnova

Contents

Preface v

1 The regularity condition. Newton's method 1
   1.1 Preliminary results 1
   1.2 Linearization procedure 2
   1.3 Error analysis 4
   Problems 6

2 The Gauss–Newton method 10
   2.1 Motivation 10
   2.2 Convergence rates 12
   Problems 14

3 The gradient method 16
   3.1 The gradient method for regular problems 16
   3.2 Ill-posed case 18
   Problems 20

4 Tikhonov's scheme 23
   4.1 The Tikhonov functional 23
   4.2 Properties of a minimizing sequence 24
   4.3 Other types of convergence 27
   4.4 Equations with noisy data 29
   Problems 30

5 Tikhonov's scheme for linear equations 32
   5.1 The main convergence result 32
   5.2 Elements of spectral theory 34
   5.3 Minimizing sequences for linear equations 35
   5.4 A priori agreement between the regularization parameter and the
        error for equations with perturbed right-hand sides 37

x   Contents

	5.5 The discrepancy principle	40
	5.6 Approximation of a quasi-solution	43
	Problems	43

6 The gradient scheme for linear equations ... 45
  6.1 The technique of spectral analysis ... 45
  6.2 A priori stopping rule ... 48
  6.3 A posteriori stopping rule ... 49
  Problems ... 53

7 Convergence rates for the approximation methods in the case of linear irregular equations ... 54
  7.1 The source-type condition (STC) ... 54
  7.2 STC for the gradient method ... 57
  7.3 The saturation phenomena ... 59
  7.4 Approximations in case of a perturbed STC ... 61
  7.5 Accuracy of the estimates ... 62
  Problems ... 63

8 Equations with a convex discrepancy functional by Tikhonov's method ... 64
  8.1 Some difficulties associated with Tikhonov's method in case of a convex discrepancy functional ... 64
  8.2 An illustrative example ... 65
  Problems ... 67

9 Iterative regularization principle ... 69
  9.1 The idea of iterative regularization ... 69
  9.2 The iteratively regularized gradient method ... 70
  Problems ... 74

10 The iteratively regularized Gauss–Newton method ... 76
  10.1 Convergence analysis ... 76
  10.2 Further properties of IRGN iterations ... 79
  10.3 A unified approach to the construction of iterative methods for irregular equations ... 83
  10.4 The reverse connection control ... 84
  Problems ... 88

11 The stable gradient method for irregular nonlinear equations ... 90
  11.1 Solving an auxiliary finite dimensional problem by the gradient descent method ... 90
  11.2 Investigation of a difference inequality ... 94
  11.3 The case of noisy data ... 95
  Problems ... 97

Contents

12 Relative computational efficiency of iteratively regularized methods . 98
   12.1 Generalized Gauss–Newton methods 98
   12.2 A more restrictive source condition 100
   12.3 Comparison to iteratively regularized gradient scheme............ 101
   Problems 102

13 Numerical investigation of two-dimensional inverse gravimetry problem.. 103
   13.1 Problem formulation 103
   13.2 The algorithm 104
   13.3 Simulations 105
   Problems 109

14 Iteratively regularized methods for inverse problem in optical tomography 111
   14.1 Statement of the problem 111
   14.2 Simple example 112
   14.3 Forward simulation 114
   14.4 The inverse problem 116
   14.5 Numerical results... 119
   Problems 121

15 Feigenbaum's universality equation 123
   15.1 The universal constants.................................... 123
   15.2 Ill-posedness 125
   15.3 Numerical algorithm for $2 \leq z \leq 12$ 125
   15.4 Regularized method for $z \geq 13$ 127
   Problems 128

16 Conclusion 130

References... 132

Index 137

1

The regularity condition. Newton's method

1.1 Preliminary results

In this textbook we primarily focus on the operator equation in the form

$$F(x) = 0, \qquad (1.1)$$

where $F : H_1 \to H_2$ is a nonlinear operator acting on a pair of real Hilbert spaces (H_1, H_2). Some results can also be generalized to the case of a complex Hilbert space. Let x^* be the solution of interest, and let $\mathcal{L}(H_1, H_2)$ be the normed space of all linear continuous operators from H_1 into H_2. Suppose that F is defined and Fréchet differentiable everywhere in H_1. Recall that $F'(x_0) \in \mathcal{L}(H_1, H_2)$ is said to be the *Fréchet derivative* of F at a point x_0 if for any x in H_1

$$F(x) - F(x_0) = F'(x_0)(x - x_0) + \omega(x_0, x),$$

$\|\omega(x_0, x)\| = o(\|x - x_0\|)$ as $x \to x_0$. Below by $\|\cdot\|_H$ and $(\cdot, \cdot)_H$ we denote the norm and the scalar product in a Hilbert space H. Throughout the book it is assumed that the following conditions on F' hold:

$$\|F'(x)\|_{\mathcal{L}(H_1, H_2)} \le N_1 \quad \forall x \in H_1; \qquad (1.2)$$
$$\|F'(x) - F'(y)\|_{\mathcal{L}(H_1, H_2)} \le N_2 \|x - y\|_{H_1} \quad \forall x, y \in H_1. \qquad (1.3)$$

For some iterative processes one may weaken conditions (1.2) and (1.3) by replacing $\forall x, y \in H_1$ with $\forall x, y \in \Omega$, where Ω is some bounded subset of H_1. One can take $\Omega = B(0, R)$, a ball of a sufficiently large radius, for example. In general,

$$B(x, r) = \{y \in H_1 : \|y - x\|_{H_1} \le r\}.$$

The replacement is usually possible when the methods under investigation are of a special nature, and it is known a priori that all iterations they generate are contained in a bounded subset of H_1. One can easily see that if we use $\forall x, y \in \Omega$ instead of $\forall x, y \in H_1$ in (1.2) and (1.3), then condition (1.2) with

$$N_1 = \|F'(\bar{x})\|_{\mathcal{L}(H_1, H_2)} + N_2 \sup_{x \in \Omega} \|x - \bar{x}\|_{H_1}$$

is a consequence of (1.3) for any $\bar{x} \in \Omega$. Therefore, inequality (1.3) alone needs to be verified in that case, and we can require the differentiability of F on Ω only. We'll keep that remark in mind, while using conditions (1.2) and (1.3) in order to simplify our presentation. Let $\mathcal{F}(N_1, N_2)$ be the class of operators F satisfying these two conditions.

It follows from (1.3) that $F'(x)$ depends continuously on x as a map from H_1 to $\mathcal{L}(H_1, H_2)$. For all operators with a continuous derivative and for operators from the class $\mathcal{F}(N_1, N_2)$ in particular, one has

$$F(x+h) - F(x) = \int_0^1 F'(x+th)h\,dt, \quad x, h \in H_1, \tag{1.4}$$

as a result of the Newton–Leibniz theorem. Clearly, from (1.3) and (1.4) one can derive the following version of Taylor's formula

$$F(x+h) = F(x) + F'(x)h + G(x,h) \quad \forall x, h \in H_1, \tag{1.5}$$

which will be used later on. Here the remainder $G(x, h)$ satisfies the estimate

$$\|G(x,h)\|_{H_2} \leq \frac{1}{2} N_2 \|h\|_{H_1}^2. \tag{1.6}$$

Finally, one more helpful inequality

$$\|F(x+h) - F(x)\|_{H_2} \leq N_1 \|h\|_{H_1} \quad \forall x, h \in H_1 \tag{1.7}$$

is also a consequence of (1.4).

1.2 Linearization procedure

One of the most popular approaches to the construction of iterative methods for various classes of equations with differentiable operators is linearization of these equations. Take an arbitrary element $x_0 \in H_1$. Equation (1.1) can be written in the form

$$F(x_0 + h) = 0, \tag{1.8}$$

where $h = x - x_0$ is a new unknown. The *linearization* procedure for equation (1.1) at the point x_0 is as follows. Discard the last term $G(x, h)$ in expression (1.5) and get the approximate identity

$$F(x+h) \approx F(x) + F'(x)h.$$

As the result, equation (1.8) becomes linear with respect to h

$$F(x_0) + F'(x_0)h = 0, \quad h \in H_1. \tag{1.9}$$

A solution to above linearized equation (1.9), if exists, defines certain element $\hat{h} \in H_1$, which is assumed to be 'an approximate solution' to equation (1.8). Hence, the point $\hat{x} = x_0 + \hat{h}$ may be considered as some kind of approximation to the solution x^* of initial equation (1.1).

Let us analyze this procedure in more detail. Obviously, it is based on the assumption that the term $G(x, h)$ in the right-hand side of (1.5) is small for $x = x_0$. Moreover, in order to deal with linear equation (1.9), to guarantee that it is uniquely solvable for any $F(x_0) \in H_2$ for example, one usually requires continuous invertibility of the operator $F'(x_0)$ together with some estimate on the norm of the inverse operator $F'(x_0)^{-1}$. Recall, that an operator $A \in \mathcal{L}(H_1, H_2)$ is said to be *continuously invertible* if there exists an inverse operator A^{-1} and $A^{-1} \in \mathcal{L}(H_2, H_1)$. Continuous invertibility of the operator $F'(x_0)$ is required due to the fact that otherwise the inverse operator $F'(x_0)^{-1}$ is either undefined, or its domain $D(F'(x_0)^{-1})$ does not coincide with the entire space H_2. As the result, equation (1.9) may be either unsolvable, or it can have infinitely many solutions. Also, if $F'(x_0)^{-1} \notin \mathcal{L}(H_2, H_1)$, then even for $F(x_0) \in D(F'(x_0)^{-1})$ the solution to (1.9) cannot continuously depend on $F(x_0)$. Therefore, small errors in the problem, which are not possible to avoid in any computational process, can change the solution considerably, or even turn (1.9) into an equation with no solutions. In other words, when $F'(x_0)$ is not continuously invertible, equation (1.9) is an *ill-posed problem*. We'll talk more on how to solve such problems below.

Let $F'(x_0)$ be continuously invertible, i.e.,

$$F'(x_0)^{-1} \in \mathcal{L}(H_2, H_1).$$

Then F is called *regular* at the point x_0. Equation (1.1) is called *regular in a neighborhood of the solution* x^*, if there is a neighborhood $B(x^*, \varepsilon)$, such that for any $x \in B(x^*, \varepsilon)$ the operator $F'(x)^{-1}$ exists and belongs to $\mathcal{L}(H_2, H_1)$. Otherwise equation (1.1) is called *irregular in a neighborhood of* x^*. Thus, irregularity of (1.1) in a neighborhood of x^* means that there exist points x, arbitrarily close to x^*, where either the operator $F'(x)^{-1}$ is undefined, or $F'(x)^{-1} \notin \mathcal{L}(H_2, H_1)$. Equation (1.1) with $F'(x)$ being a compact linear operator for all x in a neighborhood of x^*, is a typical example of irregular problem when H_1 is infinite dimensional (see Problem 1.12).

Assume that for some $m > 0$

$$\|F'(x_0)^{-1}\|_{\mathcal{L}(H_2, H_1)} \leq m < \infty. \tag{1.10}$$

Using (1.5), (1.6) (1.9), and (1.10), it is not difficult to estimate the error $\|\hat{x} - x^*\|_{H_1}$ in terms of $\|x_0 - x^*\|_{H_1}$. Indeed, from (1.9) one has

$$\hat{h} = -F'(x_0)^{-1} F(x_0),$$

and therefore

$$\hat{x} = x_0 - F'(x_0)^{-1} F(x_0).$$

Hence,

$$\|\hat{x} - x^*\|_{H_1} = \|x_0 - x^* - F'(x_0)^{-1}(F(x_0) - F(x^*))\|_{H_1}. \tag{1.11}$$

1 The regularity condition. Newton's method

By (1.5) and (1.6),
$$F(x^*) - F(x_0) = F'(x_0)(x^* - x_0) + G(x_0, x^* - x_0), \qquad (1.12)$$

where
$$\|G(x_0, x^* - x_0)\|_{H_2} \leq \frac{1}{2} N_2 \|x_0 - x^*\|_{H_1}^2.$$

Substitute (1.12) into (1.11). Then inequality (1.10) yields
$$\|\hat{x} - x^*\|_{H_1} \leq \frac{1}{2} m N_2 \|x_0 - x^*\|_{H_1}^2. \qquad (1.13)$$

From (1.13) it follows that $\hat{x}$, the solution of linearized equation (1.9), is much closer to x^*, the solution of original equation (1.1), than the initial element x_0, provided that x_0 is not too far from x^*.

The above linearization procedure can now be repeated at the point $x_1 = \hat{x}$. Thus, one gets an iterative process, which is called *Newton's method*. Formally, the process is as follows. Starting at $x_0 \in H_1$ and linearizing equation (1.1) at every step, one obtains the sequence of elements $x_0, x_1, \ldots$, where

$$x_{n+1} = x_n - F'(x_n)^{-1} F(x_n), \quad n = 0, 1, \ldots. \qquad (1.14)$$

1.3 Error analysis

Clearly, method (1.14) is well-defined only if the corresponding linear operator $F'(x_n)$ is continuously invertible at every step, i.e., $F'(x_n)^{-1} \in \mathcal{L}(H_2, H_1)$. In order to guarantee that, it is sufficient to assume that regularity condition (1.10) is fulfilled for all elements x_n ($n = 0, 1, \ldots$), defined in (1.14). However, this a posteriori assumption is not convenient in practice, since it is not possible to verify before the iterative process has begun. The following a priori assumption of uniform regularity is more suitable: inequality (1.10) holds for all $x \in H_1$, i.e.,

$$\sup_{x \in H_1} \|F'(x)^{-1}\|_{\mathcal{L}(H_2, H_1)} \leq m < \infty. \qquad (1.15)$$

For now, let us accept condition (1.15) and use it to study the behavior of iterations (1.14). Denote
$$\varepsilon_n = \|x_n - x^*\|_{H_1}.$$

By (1.13) the following estimate is satisfied:
$$\varepsilon_{n+1} \leq \frac{1}{2} m N_2 \varepsilon_n^2, \quad n = 0, 1, \ldots. \qquad (1.16)$$

1.3 Error analysis

With (1.16) one can get an explicit estimate on ε_n in terms of ε_0. To this end, one has to estimate ε_1 in terms of ε_0, ε_2 in terms of ε_1, etc. Finally, one arrives at the inequality

$$\varepsilon_n \leq \frac{2}{mN_2} \left(\frac{1}{2} m N_2 \varepsilon_0 \right)^{2^n}, \quad n = 0, 1, \ldots. \tag{1.17}$$

Based on (1.17), one concludes that if

$$\|x_0 - x^*\|_{H_1} \leq \frac{2q}{mN_2}, \quad q \in (0, 1), \tag{1.18}$$

then

$$\varepsilon_n \leq \frac{2}{mN_2} q^{2^n}, \quad n = 0, 1, \ldots,$$

and the sequence $\{\varepsilon_n\}$ converges to zero rapidly:

$$\lim_{n \to \infty} \varepsilon_n = \lim_{n \to \infty} \|x_n - x^*\|_{H_1} = 0.$$

Keeping in mind the relationship $\varepsilon_{n+1} = O(\varepsilon_n^2)$, see (1.16), we say that the sequence $\{\varepsilon_n\}$ converges to zero at a quadratic rate.

Inequality (1.18) implies that the initial approximation x_0 must be sufficiently close to the solution of interest x^*. Even in case $H_1 = \mathbb{R}$, when F in (1.1) is a function in one variable, there are examples of divergent sequences for initial approximations that are distant from the solution (see Problem 1.3). Iterative methods of this kind are said to be locally convergent. The above remark does not mean that iterations (1.14) always diverge if condition (1.18) is violated. There are trivial examples of finite dimensional equations and systems of equations, for which Newton's method converges with any choice of x_0 (see Problems 1.4 and 1.6). Assumption of uniform regularity (1.15) is not necessary for the convergence as well (see Problem 1.6).

Estimate (1.18) yields

$$\frac{1}{2} m N_2 \varepsilon_0 \leq \frac{1}{2} m N_2 \frac{2q}{mN_2} < 1.$$

Hence, it follows from (1.17) that for any $n \in \mathbb{N}$

$$\varepsilon_n \leq \frac{2}{mN_2} \left(\frac{1}{2} m N_2 \varepsilon_0 \right)^{2^n} \leq \frac{2}{mN_2} \frac{1}{2} m N_2 \varepsilon_0 = \varepsilon_0.$$

Thus, for the convergence of Newton's method it is sufficient for (1.15) to be fulfilled in some neighborhood $B(x^*, R)$ of the solution x^*, and not in the entire space H_1:

$$\sup_{x \in B(x^*, R)} \|F'(x)^{-1}\|_{\mathcal{L}(H_2, H_1)} \leq m < \infty. \tag{1.19}$$

In this case, one has to guarantee that all points x_0, satisfying (1.18), are contained in the ball $B(x^*, R)$. Obviously, this is true if

$$\frac{2q}{mN_2} \le R.$$

Finally, taking into account the remark from Section 1.1, we can replace (1.2) and (1.3) with one condition

$$\|F'(x) - F'(y)\|_{\mathcal{L}(H_1, H_2)} \le N_2 \|x - y\|_{H_1} \quad \forall x, y \in B(x^*, R). \tag{1.20}$$

Unfortunately, in the above requirements, sufficient for the convergence of Newton's scheme, it is assumed that we have prior knowledge of the location of x^*. That is a disadvantage. A more accurate convergence analysis (see [48], ch. XVIII) allows one to get the following sufficient conditions:

$$\|F'(x_0)^{-1}\|_{\mathcal{L}(H_2, H_1)} \le m_0, \quad \|F'(x_0)^{-1} F(x_0)\|_{H_1} \le \eta_0, \tag{1.21}$$

which only have to do with the initial point x_0. The precise statement is as follows. Let inequality (1.20) hold for all $x, y \in B(x_0, R)$, and the constants m_0, η_0 in (1.21) satisfy the estimates:

$$\eta_0 < \frac{1}{2 m_0 N_2}, \quad \frac{1 - \sqrt{1 - 2 m_0 \eta_0 N_2}}{m_0 N_2} \le R. \tag{1.22}$$

Then the set $\{x \in B(x_0, R) : F(x) = 0\}$ consists of a unique point x^*; iterations (1.14) are well-defined for all $n \in \mathbb{N}$, and

$$\varepsilon_n = \|x_n - x^*\|_{H_1} \le \frac{1}{m_0 N_2} (2 m_0 \eta_0 N_2)^{2^n}, \quad n = 0, 1, \ldots.$$

The last estimate is similar to (1.17). One can easily see that for m_0, N_2, and R fixed, both inequalities in (1.22) are automatically fulfilled if η_0 is sufficiently small. Note, by (1.14) and (1.21), η_0 is an upper bound for $\|x_1 - x_0\|$, i.e., for the the magnitude of the displacement at the first step of the process.

In conclusion, we would like to point out that even with these new sufficient conditions, the implementation of Newton's method in irregular case remains problematic. Although condition (1.22) does not require any a priori information about x^*, it is often hard to verify, because determination of the point x_0, satisfying (1.21) and (1.22), may be as difficult as solving (1.1) itself.

Problems

1.1. Prove formula (1.4) and estimates (1.6), (1.7).

1.2. Prove estimate (1.17).

1.3. Give an example of the equation $f(x) = 0$ ($x \in \mathbb{R}$), solvable, with a continuously differentiable function f, for which Newton's method

$$x_{n+1} = x_n - \frac{f(x_n)}{f'(x_n)}, \quad n = 0, 1, \ldots \tag{1.23}$$

fails to converge for some $x_0 \in \mathbb{R}$. Show that this can happen even if f is in the class $\mathcal{F}(N_1, N_2)$ and satisfies condition (1.15): $|f'(x)| \geq m^{-1} > 0 \ \forall x \in \mathbb{R}$. Hint: use geometrical interpretation of process (1.23).

1.4. Investigate the convergence of Newton's method (1.23) for the equations

$$x^3 - x = 0, \quad e^x - 1 = 0, \quad \sqrt{x^2 + 1} - 2 = 0,$$

depending on the choice of initial point x_0. When is the method not executable? Which of the three functions are in the class $\mathcal{F}(N_1, N_2)$? Give an example of a function $f \in \mathcal{F}(N_1, N_2)$, such that (1.23) converges to a solution of $f(x) = 0$ ($x \in \mathbb{R}$) for any initial approximation.

1.5. Let the operator $F : \mathbb{R}^n \to \mathbb{R}^n$ take the form

$$F(x^{(1)}, x^{(2)}, \ldots, x^{(n)}) = \begin{pmatrix} f_1(x^{(1)}, x^{(2)}, \ldots, x^{(n)}) \\ \vdots \\ f_n(x^{(1)}, x^{(2)}, \ldots, x^{(n)}) \end{pmatrix}.$$

Write iterative process (1.14) for the system of equations $F(x) = 0$. Give your answer in a form that does not contain inverse matrices.

1.6. Let $H_1 = H_2 = \mathbb{R}^2$ and the operator norm in $\mathcal{L}(H_2, H_1)$ be induced by the Euclidean vector norm. Analyze the convergence of iterations (1.14) for the following systems of equations

$$\begin{cases} x^{(1)} - a = 0 \\ (x^{(1)})^2 + x^{(2)} - b = 0 \end{cases}, \quad \begin{cases} x^{(1)} + \frac{1}{3}(x^{(1)} + x^{(2)})^3 - a = 0 \\ x^{(2)} - \frac{1}{3}(x^{(1)} + x^{(2)})^3 - b = 0 \end{cases}.$$

Does condition of uniform regularity (1.15) hold for the above systems? Are the operators of these systems from the class $\mathcal{F}(N_1, N_2)$? Check whether conditions (1.19) and (1.20) are satisfied here, and if so then find the constants $m = m(R)$, $N_2 = N_2(R)$.

1.7. Assume that in the first system of Problem 1.6, $a = 0, b = 1$. Find the constants $m = m(R), N_2 = N_2(R)$, which ensure (1.19) and (1.20). What set of convergence is ensured by inequality (1.18) for iterations (1.14) in that case? How does this set depend on R?

1.8. Let $F \in \mathcal{F}(N_1, N_2)$ and

$$\|F'(x^*)^{-1}\|_{\mathcal{L}(H_2, H_1)} \leq m.$$

Show that there exist a neighborhood $B(x^*, r)$ and a constant $m_1 = m_1(r)$, such that

$$\|F'(x)^{-1}\|_{\mathcal{L}(H_2, H_1)} \leq m_1 \quad \forall x \in B(x^*, r).$$

1.9. Investigate whether conditions (1.21) and (1.22) are fulfilled for the equation $x^3 - 1 = 0$. Are they satisfied at the point $x_0 = -2^{-1/3}$? How do iterations (1.23) behave if one starts at this initial point?

1.10. Let a function $K(t, s)$ be continuous on $[a, b] \times [a, b]$, and functions $g(s, \sigma)$, $\frac{\partial^2 g}{\partial \sigma^2}(s, \sigma)$ be continuous on $[a, b] \times \mathbb{R}$. Let also $f \in L_2(a, b)$ and $\lambda \in \mathbb{R}$. Suppose

$$\sup_{s \in [a,b], \sigma \in \mathbb{R}} \left| \frac{\partial^2 g}{\partial \sigma^2}(s, \sigma) \right| < \infty.$$

Define the operator $F : L_2(a, b) \to L_2(a, b)$ as

$$[F(x)](t) = x(t) - \lambda \int_a^b K(t, s) g(s, x(s)) \, ds - f(t), \quad t \in (a, b). \quad (1.24)$$

Show that the operator F, $F : H_1 = L_2(a, b) \to H_2 = L_2(a, b)$, is Fréchet differentiable and its derivative is given by the formula

$$[F'(x)h](t) = h(t) - \lambda \int_a^b K(t, s) \frac{\partial g}{\partial \sigma}(s, x(s)) h(s) \, ds, \quad h \in L_2(a, b). \quad (1.25)$$

What additional requirements on the function $g(s, \sigma)$ will guarantee that $F \in \mathcal{F}(N_1, N_2)$? Verify, that if

$$\sup_{s \in [a,b], \sigma \in \mathbb{R}} \left| \frac{\partial g}{\partial \sigma}(s, \sigma) \right| < \infty$$

and $|\lambda| \leq \epsilon$, where ϵ is sufficiently small, then operator (1.24) satisfies condition (1.15). *Hint*: it is enough to check that the norm of the solution to the linear equation $F'(x)h = z$ ($\|z\|_{H_2} = 1$) can be estimated by a constant independent on x and z. One can use the contraction mapping principle in order to get this estimate.

1.11. Use the result of Problem 1.10 to write down iterative process (1.14) for solving the equation

$$x(t) - \lambda \int_a^b K(t, s) g(s, x(s)) \, ds = f(t), \quad t \in (a, b). \quad (1.26)$$

Present the iterations in a form that does not contain inverse operators.

1.12. Answer the questions of Problem 1.10 for the operator

$$[F(x)](t) = \int_a^b K(t, s) g(s, x(s)) \, ds - f(t), \quad t \in (a, b). \quad (1.27)$$

Show that $F'(x)$ is a compact operator for any $x \in L_2(a,b)$. Recall, that a bounded linear operator $A : H_1 \to H_2$ is *compact* if and only if the image of any bounded set is a relatively compact set, i.e., if the closure of this image is a compact subset of H_2.

1.13. Generalize the results of Problems 1.10–1.12 to the case of the following operators

$$[F(x)](t) = x(t) - \lambda \int_a^b \varphi(t, s, x(s))\, ds - f(t), \tag{1.28}$$

$$[F(x)](t) = \int_a^b \varphi(t, s, x(s))\, ds - f(t), \quad t \in (a, b). \tag{1.29}$$

Prove that if the function $\varphi = \varphi(t, s, \sigma)$ is twice continuously differentiable with respect to σ for $t, s \in [a, b]$, $\sigma \in \mathbb{R}$, then condition

$$\sup_{s \in [a,b], \sigma \in \mathbb{R}} \left| \frac{\partial^2 \varphi}{\partial \sigma^2}(t, s, \sigma) \right| < \infty$$

implies (1.3) in both cases.

2

The Gauss–Newton method

2.1 Motivation

The Gauss–Newton method is more applicable than the Newton one, especially if the spaces H_1 and H_2 are not the same. Consider the following functional $\Phi : H_1 \to \mathbb{R}$

$$\Phi(x) = \frac{1}{2}\|F(x)\|_{H_2}^2, \quad x \in H_1. \tag{2.1}$$

Obviously, this functional is nonnegative and, therefore, it is bounded from below on H_1. Assume, Φ has a global minimizer x^*, which is not necessarily unique, i.e., there exists

$$x^* \in H_1, \quad \Phi(x^*) = \inf_{x \in H_1} \Phi(x). \tag{2.2}$$

In a particular case, when the set of the solutions to equation (1.1) is nonempty, the infimum in (2.2) is attained exactly at this set, and

$$\inf_{x \in H_1} \Phi(x) = 0.$$

If one has

$$\inf_{x \in H_1} \Phi(x) > 0,$$

then any element x^*, satisfying condition (2.2), can be viewed as some generalization of a solution to (1.1). Such an element is called a *least squares solution* (or *quasisolution*) to (1.1).

Take arbitrary $x_0 \in H_1$. For all points x, sufficiently close to x_0,

$$F(x) \approx F(x_0) + F'(x_0)(x - x_0).$$

Using the above approximation, replace the operator F in the right-hand side of (2.1) by the following affine operator

$$F(x_0) + F'(x_0)(x - x_0).$$

As the result of this substitution, one arrives at the quadratic functional

$$\Phi(x_0; x) = \frac{1}{2} \|F(x_0) + F'(x_0)(x - x_0)\|_{H_2}^2, \quad x \in H_1.$$

This functional is convex with respect to x, therefore the set of its minimizers in H_1 coincides with the solution set of the following equation [28]

$$\Phi'_x(x_0; x) = 0.$$

Here $\Phi'_x(x_0; x)$ is the gradient of the functional $\Phi = \Phi(x_0; x)$ with respect to x. The direct calculation yields

$$\Phi'_x(x_0; x) = F'^*(x_0)[F(x_0) + F'(x_0)(x - x_0)]. \tag{2.3}$$

Hence we get a linear operator equation in the space H_1:

$$F'^*(x_0) F(x_0) + F'^*(x_0) F'(x_0)(x - x_0) = 0, \quad x \in H_1. \tag{2.4}$$

In (2.4) and further, A^* is a linear operator conjugate to $A \in \mathcal{L}(H_1, H_2)$. It is known that $A^* \in \mathcal{L}(H_2, H_1)$. Equation (2.4) can also be derived from (1.9), if one apples the operator $F'^*(x_0)$ to both sides of (1.9).

Let us point out, that F belonging to the class $\mathcal{F}(N_1, N_2)$ does not necessarily guarantee the existence of a minimizer for the functional $\Phi = \Phi(x_0; x)$. Thus, solvability of (2.4) is not guaranteed either. Equation (2.4) is solvable if, for example, the operator $F'^*(x_0) F'(x_0)$ is continuously invertible. For this reason, it is appropriate to extend the notion of regularity of the operator F at a point x_0 introduced in Chapter 1, i.e., to refer to the operator as *regular* at the point x_0 not only if

$$F'(x_0)^{-1} \in \mathcal{L}(H_2, H_1),$$

but also if

$$(F'^*(x_0) F'(x_0))^{-1} \in \mathcal{L}(H_1, H_1).$$

In accordance with that, we now call equation (1.1) *regular in a neighborhood of the solution* x^*, if there exists a neighborhood $B(x^*, \varepsilon)$, such that

$$F'(x)^{-1} \in \mathcal{L}(H_2, H_1) \quad \forall x \in B(x^*, \varepsilon),$$

or

$$(F'^*(x) F'(x))^{-1} \in \mathcal{L}(H_1, H_1) \quad \forall x \in B(x^*, \varepsilon).$$

Whenever F is regular at the point x_0, equation (2.4) is uniquely solvable, and its solution $\hat{x}$ is given by the formula

$$\hat{x} = x_0 - (F'^*(x_0) F'(x_0))^{-1} F'^*(x_0) F(x_0).$$

If the following condition of uniform regularity is satisfied

$$\sup_{x \in H_1} \|(F'^*(x)F'(x))^{-1}\|_{\mathcal{L}(H_1,H_1)} \leq m < \infty, \tag{2.5}$$

then at the point $x_1 = \hat{x}$ one can consider the functional $\Phi = \Phi(x_1; x)$, get the equation similar to (2.4), solve it, consider $\Phi = \Phi(x_2; x)$, etc. As the result, one obtains the sequence of approximations $x_0, x_1, \ldots$, where

$$x_{n+1} = x_n - (F'^*(x_n)F'(x_n))^{-1}F'^*(x_n)F(x_n), \quad n = 0, 1, \ldots. \tag{2.6}$$

Iterative process (2.6) is called *the Gauss–Newton method*.

2.2 Convergence rates

Let us study the possibility of convergence of iterations (2.6) to the element x^*, a global minimizer of functional (2.1) in the space H_1. Note, that at the point x^* the necessary condition of minimum

$$\Phi'(x^*) = 0$$

must be fulfilled. This condition, together with the identity

$$\Phi'(x) = F'^*(x)F(x), \quad x \in H_1, \tag{2.7}$$

imply

$$F'^*(x^*)F(x^*) = 0. \tag{2.8}$$

Hence, by (2.6) and (2.8), one concludes

$$\begin{aligned}
x_{n+1} - x^* &= x_n - x^* \\
&\quad - (F'^*(x_n)F'(x_n))^{-1}[F'^*(x_n)F(x_n) \\
&\quad - F'^*(x_n)F(x^*) + F'^*(x_n)F(x^*) - F'^*(x^*)F(x^*)] \\
&= \{x_n - x^* - (F'^*(x_n)F'(x_n))^{-1}F'^*(x_n)(F(x_n) - F(x^*))\} \\
&\quad - \{(F'^*(x_n)F'(x_n))^{-1}(F'^*(x_n) - F'^*(x^*))F(x^*)\} = \mathcal{A}_n - \mathcal{B}_n.
\end{aligned}$$

Applying inequalities (1.2) and (1.6), one estimates the norm of $\mathcal{A}_n$ as follows

$$\|\mathcal{A}_n\|_{H_1} \leq \frac{1}{2} m N_1 N_2 \|x_n - x^*\|_{H_1}^2. \tag{2.9}$$

It is easy to see, that inequality (2.9) is analogous to (1.13) in Section 1.2. The estimate for $\|\mathcal{B}_n\|_{H_1}$ is the consequence of (1.3) and (2.5)

$$\|\mathcal{B}_n\|_{H_1} \leq m N_2 \|F(x^*)\|_{H_2} \|x_n - x^*\|_{H_1}. \tag{2.10}$$

To derive (2.9) and (2.10), we use the following property of a conjugate operator:

$$\|B^*\|_{\mathcal{L}(H_2,H_1)} = \|B\|_{\mathcal{L}(H_1,H_2)} \quad \forall B \in \mathcal{L}(H_1, H_2). \tag{2.11}$$

The primary case, that will interest us in the future, is the case of zero minimum of functional (2.1), when

$$\Phi(x^*) = \inf_{x \in H_1} \Phi(x) = 0.$$

Thus, for

$$\varepsilon_n = \|x_n - x^*\|_{H_1},$$

one gets

$$\varepsilon_{n+1} \leq \frac{1}{2} m N_1 N_2 \varepsilon_n^2 + m N_2 \|F(x^*)\|_{H_2} \varepsilon_n, \quad n = 0, 1, \ldots.$$

Then $F(x^*) = 0$, and the above estimate for ε_{n+1} becomes similar to (1.16). However, the meaning of the constant m in the right-hand side is different. Using the same argument as in Chapter 1, we verify that if

$$\|x_0 - x^*\|_{H_1} \leq \frac{2q}{m N_1 N_2}, \quad q \in (0, 1), \tag{2.12}$$

then the sequence $\{\varepsilon_n\}$ converges to zero at a quadratic rate, and, therefore, the sequence $\{x_n\}$ converges (as $n \to \infty$) to the solution x^* of equation (1.1) in the norm of H_1. Also, as in Chapter 1, one can show that conditions (1.2) and (1.3) may be replaced with (1.20), if q in (2.12) satisfies the inequality

$$\frac{2q}{m N_1 N_2} \leq R. \tag{2.13}$$

Finally, the uniform regularity condition can be taken in the form

$$\sup_{x \in B(x^*, R)} \|(F'^*(x) F'(x))^{-1}\|_{\mathcal{L}(H_1, H_1)} \leq m. \tag{2.14}$$

In general, when $\|F(x^*)\|_{H_2} > 0$, conditions (1.20) and (2.14) do not imply even the local convergence of the Gauss–Newton process. The local convergence of $\{x_n\}$ to x^* at a linear rate can only be guaranteed if $\|F(x^*)\|_{H_2}$ is small enough (see Problem 2.4). The different nature of convergence can be explained as follows. According to (2.7), variational problem (2.2) is formally equivalent to the operator equation

$$\mathcal{F}(x) = 0, \quad x \in H_1, \tag{2.15}$$

where $\mathcal{F}(x) = \Phi'(x) = F'^*(x) F(x)$. Equation (2.15) can be solved by Newton's method. Assuming that F is twice Fréchet differentiable, one has

$$\mathcal{F}'(x)h = F'^*(x) F'(x)h + (F''(x)h)^* F(x), \quad h \in H_1. \tag{2.16}$$

However, the iterative process

$$x_{n+1} = x_n - \mathcal{F}'(x_n)^{-1}\mathcal{F}(x_n), \quad n = 0, 1, \ldots, \qquad (2.17)$$

is inconvenient in practice, since its implementation requires the second derivative of the operator F. One can resolve this difficulty by omitting the second term in the right-hand side of (2.16). As the result, one gets the Gauss–Newton method. Thus, process (2.6) can be viewed as a simplified version of (1.14). The changes made to (2.17) are minimal for $F(x^*) = 0$, and this is the case when the Gauss–Newton method converges quadratically. If $F(x^*) \neq 0$, then the error introduced by omitting the second term is greater, and it gets bigger as $\|F(x^*)\|_{H_2}$ is increasing. Hence, the convergence rate of (2.6) is reduced from quadratic to linear. For large values of $\|F(x^*)\|_{H_2}$ the method fails to converge.

Problems

2.1. Prove identities (2.3) and (2.7).

2.2. Show that the functional $\Phi = \Phi(x_0; x)$ is convex with respect to x.

2.3. Derive estimates (2.9) and (2.10).

2.4. Let conditions (1.20) and (2.14) hold, and

$$m N_2 \|F(x^*)\|_{H_2} = \tilde{q} < 1.$$

Prove that if

$$\varepsilon_0 \leq \min\left\{\frac{2(q - \tilde{q})}{m N_1 N_2}, R\right\}, \quad q \in (\tilde{q}, 1),$$

then $\lim_{n \to \infty} \varepsilon_n = 0$, and the following estimate is satisfied

$$\varepsilon_n \leq q^n \varepsilon_0, \quad n = 0, 1, \ldots.$$

2.5. Let the operator $F : \mathbb{R}^n \to \mathbb{R}^m$ take the form $F(x) = Ax - f$, where A is an $m \times n$ matrix and $b \in \mathbb{R}^m$. Show that under these assumptions the regularity condition is satisfied if and only if the columns of A are linearly independent. What relation between m and n should necessarily be fulfilled for this condition to hold? How can this necessary condition be generalized to the case of arbitrary nonlinear map $F : \mathbb{R}^n \to \mathbb{R}^m$? Is Newton's method (1.14) always executable in that case?

2.6. Investigate the applicability of iterations (2.6) to the following systems of equations

$$\begin{cases} x^{(1)} = 0 \\ x^{(1)} - x^{(2)} = 0 \\ (x^{(1)})^2 - x^{(1)} - x^{(2)} = 0 \end{cases}, \quad \begin{cases} (x^{(1)})^2 - (x^{(2)})^2 = 0 \\ x^{(1)} x^{(2)} = 0 \end{cases}.$$

What set of convergence is guaranteed by inequality (2.12) and conditions (1.20), (2.13), (2.14) for the first system?

2.7. Show that under the assumptions of Problem 1.10 the operator, conjugate to (1.25), is as follows

$$[F'^*(x)h](t) = h(t) - \lambda \frac{\partial g}{\partial \sigma}(t, x(t)) \int_a^b K(s,t) h(s)\, ds, \quad h \in L_2(a,b). \quad (2.18)$$

Use the above formula to write method (2.6) for equation (1.26) in the form that does not contain inverse operators. Prove that the iterations converge, if $|\lambda|$ is rather small.

2.8. Check whether the Gauss–Newton method is executable for the equations with operators (1.27) and (1.29).

2.9. Write the Gauss–Newton method for the equation with operator (1.28).

3

The gradient method

3.1 The gradient method for regular problems

In this Chapter we consider the *gradient method*, which is used to iteratively approximate a solution of equation (1.1), or, more precisely, a minimizer of functional (2.1). Following Chapter 2, suppose that x^* is a solution to (1.1) and, therefore, functional (2.1) attains global zero minimum at x^*. The method under investigation is an iterative procedure

$$x_{n+1} = x_n - \gamma \Phi'(x_n) = x_n - \gamma F'^*(x_n) F(x_n), \quad n = 0, 1, \ldots, \quad x_0 \in H_1. \tag{3.1}$$

The scalar $\gamma > 0$ in (3.1) is called a step size. If at each iteration n, γ is is chosen to minimize $\tilde{\Phi}(\gamma) = \Phi(x_n - \gamma \Phi'(x_n))$, then one gets the *steepest descent method*. If one fixes γ with $0 < \gamma < \frac{1}{N_1^2}$, one gets a method known as *Landweber iteration* (see [80], ch. I).

Let us analyze conditions, sufficient for the convergence of sequence (3.1) to the point x^*. Formula (3.1), our assumption on x^*, and Taylor's expansion (1.5) imply

$$x_{n+1} - x^* = x_n - x^* - \gamma F'^*(x_n)(F(x_n) - F(x^*))$$
$$= [E_1 - \gamma F'^*(x_n) F'(x_n)](x_n - x^*) + \gamma F'^*(x_n) G(x_n, x^* - x_n).$$

Here and below E_j is an identity operator in the space H_j, $j = 1, 2$. Keeping our previous notation ε_n for $\|x_n - x^*\|_{H_1}$, using (2.11) and inequalities (1.2), (1.6), we obtain the following estimate

$$\varepsilon_{n+1} \leq \|E_1 - \gamma F'^*(x_n) F'(x_n)\|_{\mathcal{L}(H_1, H_1)} \varepsilon_n + \frac{1}{2} \gamma N_1 N_2 \varepsilon_n^2, \quad n = 0, 1, \ldots. \tag{3.2}$$

The linear operator

$$F'^*(x_n) F'(x_n) \in \mathcal{L}(H_1, H_1)$$

is self-adjoint in the space H_1. Condition (1.2) yields

$$\|F'^*(x_n)F'(x_n)\|_{\mathcal{L}(H_1,H_1)} \leq N_1^2.$$

Hence the spectrum $\sigma(F'^*(x_n)F'(x_n))$ lies on the segment $[l_n, L_n]$ of the real axis and for any n

$$l_n \geq 0, \quad L_n \leq N_1^2.$$

In this textbook, by $\sigma(B)$ we always denote the spectrum of an operator $B \in \mathcal{L}(H_1, H_1)$. By the spectral theory of self-adjoint operators [28]

$$\|E_1 - \gamma F'^*(x_n)F'(x_n)\|_{\mathcal{L}(H_1,H_1)} = \max_{\lambda \in \sigma(F'^*(x_n)F'(x_n))} |1 - \gamma\lambda|$$

$$\leq \max_{\lambda \in [l_n, L_n]} |1 - \gamma\lambda| = q_n(\gamma). \quad (3.3)$$

From estimates (3.2) and (3.3) one concludes, that convergence of ε_n to zero (as $n \to 0$) is guaranteed only if

$$q_n(\gamma) \leq q < 1.$$

The last condition holds for a sufficiently small $\gamma > 0$, if the spectra of all operators $F'^*(x)F'(x)$ are separated from zero, uniformly with respect to $x \in H_1$, i.e.,

$$\inf\{\lambda : \lambda \in \sigma(F'^*(x)F'(x)), x \in H_1\} = l > 0. \quad (3.4)$$

In particular, it follows from (3.4) that

$$l_n \geq l > 0, \quad n = 0, 1, \ldots. \quad (3.5)$$

By (3.3) and (3.5),

$$q_n(\gamma) \leq \max_{\lambda \in [l, N_1^2]} |1 - \gamma\lambda| = q(\gamma), \quad n = 0, 1, \ldots.$$

It is easy to see, that for every γ fixed, $\max_{\lambda \in [l, N_1^2]} |1 - \gamma\lambda|$ is attained at one of the endpoints of $[l, N_1^2]$, i.e.,

$$q(\gamma) = \max\{|1 - \gamma l|, |1 - \gamma N_1^2|\}. \quad (3.6)$$

Choosing the step size $\gamma > 0$ rather small, one can make both constants, $|1 - \gamma l|$ and $|1 - \gamma N_1^2|$, strictly less than 1. Thus, by (3.2), (3.3), (3.6), for a small positive γ, one has

$$\varepsilon_{n+1} \leq q(\gamma)\varepsilon_n + \frac{1}{2}\gamma N_1 N_2 \varepsilon_n^2, \quad n = 0, 1, \ldots, \quad (3.7)$$

The convergence analysis of iterative process (3.1) can now be done based on the result of Problem 2.4. This result, in particular, indicates that condition (3.4) may be replaced with a weaker one

$$\inf\{\lambda : \lambda \in \sigma(F'^*(x)F'(x)), x \in B(x^*, R)\} = l > 0, \quad (3.8)$$

which is equivalent to the condition of uniform regularity (2.14). Using (3.8) and arguing like in Problem 2.4, we set

$$\varepsilon_0 \leq \min\left\{\frac{2(q_1 - q(\gamma))}{\gamma N_1 N_2}, R\right\}, \quad q_1 \in (q(\gamma), 1),$$

and obtain the following estimate

$$\varepsilon_n \leq q_1^n \varepsilon_0, \quad n = 0, 1, \ldots,$$

from (3.7). Thus, the uniform regularity condition implies local convergence of the gradient process to a solution of equation (1.1) at a linear rate. The Newton and the Gauss–Newton methods are both quadratically convergent, if x^* is a solution to (1.1). On the other hand, practical implementation of process (3.1) is easier as compared to the Newton and the Gauss–Newton ones.

3.2 Ill-posed case

If the operator F in equation (1.1) does not satisfy regularity condition (2.14) (or (3.8)), then it is not possible to guarantee even local convergence of process (3.1), although, unlike the Newton and the Gauss–Newton schemes, the gradient method is formally executable for any differentiable operator F. It turns out, however, that process (3.1) has some interesting properties even when the regularity condition is violated (ill-posed case). Let us consider the behavior of the sequence $\{\Phi(x_n)\}$ in more detail, without assuming (3.4) or (3.8). One can prove that, for a special choice of the step size γ, this sequence is monotonically decreasing. Moreover, it emerges that

$$\lim_{n \to \infty} \|\Phi'(x_n)\|_{H_1} = 0. \tag{3.9}$$

To verify (3.9), we use the following auxiliary results.

1) Let $x, y \in H_1$. Consider the scalar function $\Phi(x + ty)$ of a parameter $t \in [0, 1]$. Since $F \in \mathcal{F}(N_1, N_2)$, this function is differentiable with respect to t, and the Newton–Leibniz formula holds

$$\Phi(x + y) - \Phi(x) = \int_0^1 \frac{d}{dt} \Phi(x + ty)\, dt = \int_0^1 (\Phi'(x + ty), y)_{H_1}\, dt.$$

The above identity can be written as

$$\Phi(x + y) = \Phi(x) + (\Phi'(x), y)_{H_1} + \int_0^1 (\Phi'(x + ty) - \Phi'(x), y)_{H_1}\, dt. \tag{3.10}$$

2) For any $x, y \in H_1$ the following estimate is fulfilled

$$\|\Phi'(x) - \Phi'(y)\|_{H_1} \leq (N_2 \min\{\|F(x)\|_{H_2}, \|F(y)\|_{H_2}\} + N_1^2)\, \|x - y\|_{H_1}. \tag{3.11}$$

Indeed, by (1.2), (1.3), (1.7), and (2.11),

$$\begin{aligned}\|\Phi'(x) - \Phi'(y)\|_{H_1} &= \|F'^*(x)F(x) - F'^*(y)F(y)\|_{H_1} \\ &\leq \|F'^*(x)F(x) - F'^*(y)F(x)\|_{H_1} + \|F'^*(y)F(x) - F'^*(y)F(y)\|_{H_1} \\ &\leq (N_2\|F(x)\|_{H_2} + N_1^2)\|x - y\|_{H_1}. \end{aligned} \quad (3.12)$$

In the above argument, the variables x and y can be interchanged. As the result, $\|F(x)\|_{H_2}$ is replaced with $\|F(y)\|_{H_2}$ in the right-hand side of (3.12), and estimate (3.11) is derived.

Let in (3.10)

$$x = x_0, \quad y = -\gamma \Phi'(x_0).$$

By (3.1) $x + y = x_1$, therefore

$$\Phi(x_1) = \Phi(x_0) - \gamma \|\Phi'(x_0)\|_{H_1}^2 \\ - \gamma \int_0^1 (\Phi'(x_0 - t\gamma\Phi'(x_0)) - \Phi'(x_0), \Phi'(x_0))_{H_1} \, dt. \quad (3.13)$$

From (3.12) one concludes

$$\|\Phi'(x_0 - t\gamma\Phi'(x_0)) - \Phi'(x_0)\|_{H_1} \\ \leq (N_2\|F(x_0)\|_{H_2} + N_1^2)t\gamma\|\Phi'(x_0)\|_{H_1} = Lt\gamma\|\Phi'(x_0)\|_{H_1}, \quad (3.14)$$

where

$$L = N_2\|F(x_0)\|_{H_2} + N_1^2. \quad (3.15)$$

Applying inequality (3.14) to estimate the right-hand side of (3.13), one gets

$$\Phi(x_1) \leq \Phi(x_0) - \gamma\|\Phi'(x_0)\|_{H_1}^2 + \gamma^2 L \|\Phi'(x_0)\|_{H_1}^2 \int_0^1 t \, dt \\ = \Phi(x_0) - \gamma\left(1 - \frac{1}{2}\gamma L\right)\|\Phi'(x_0)\|_{H_1}^2. \quad (3.16)$$

Take the step size γ, such that

$$0 < \gamma < \frac{2}{L}. \quad (3.17)$$

Then, from (3.16) and (3.17), it follows that

$$\Phi(x_1) \leq \Phi(x_0), \quad (3.18)$$

where inequality (3.18) is strict, if $\Phi'(x_0) \neq 0$, i.e., if the necessary condition of minimum of the functional Φ is not satisfied at x_0. Note, that by (3.18),

$$\|F(x_1)\|_{H_2} \leq \|F(x_0)\|_{H_2}. \quad (3.19)$$

Now one can repeat the above argument for x_1. Obviously, if one replaces x_0 with x_1 in the right-hand side of (3.16), and x_1 with x_2 in the left-hand side, inequality (3.16) still holds with the same constant L, defined in (3.15). That follows from inequality (3.19). Hence, if one defines

$$\beta = \gamma\left(1 - \frac{1}{2}\gamma L\right),$$

then one has

$$\Phi(x_{n+1}) \leq \Phi(x_n) - \beta \|\Phi'(x_n)\|_{H_1}^2, \quad n = 0, 1, \ldots. \tag{3.20}$$

Sum up estimates (3.20) with respect to n from $n = 0$ to $n = M$, then

$$\sum_{n=0}^{M} \|\Phi'(x_n)\|_{H_1}^2 \leq \frac{\Phi(x_0) - \Phi(x_{M+1})}{\beta} \leq \frac{\Phi(x_0)}{\beta}. \tag{3.21}$$

Since $M \in \mathbb{N}$ is arbitrary, (3.21) implies that the series

$$\sum_{n=0}^{\infty} \|\Phi'(x_n)\|_{H_1}^2$$

converges, and, therefore, identity (3.9) is true. Also, one can conclude from (3.20), that the sequence $\{\Phi(x_n)\}$ is monotonically decreasing ($\Phi(x_{n+1}) \leq \Phi(x_n)$, $n = 0, 1, \ldots$). This result is rather weak, because one can say nothing about the convergence of $\{x_n\}$ itself, as $n \to \infty$. Moreover, even if x^* is a global minimizer of functional (2.1), and $\Phi(x^*) = 0$, one cannot assert that

$$\lim_{n \to \infty} \Phi(x_n) = 0, \tag{3.22}$$

i.e., that sequence (3.1) is minimizing.

Unfortunately, that is all we get for functional (2.1), when it is generated by an irregular operator $F \in \mathcal{F}(N_1, N_2)$. One fails to be more specific about iterations (3.1) in this case. A major obstacle here is that we do not assume the functional Φ to be convex. Thus, already for $H_1 = \mathbb{R}$, even if $\Phi(x^*) = 0$ and (3.22) holds, sequence (3.1) does not necessarily converge to the solution x^* (see Problem 3.4). On the other hand, convexity of the functional Φ does not guarantee convergence of x_n to x^* in the norm of H_1 either (see Problem 3.5). The most favorable situation to investigate convergence of gradient-type methods is the one of a strongly convex functional Φ (see Problem 3.7). In this case, the infimum in (2.2) is attained at a unique point, and for most gradient-type methods convergence is proven with any initial approximation (global convergence). However, if equation (1.1) is irregular, functional (2.1) is not strongly convex.

Problems

3.1. Find minimum of function (3.6) for $\gamma > 0$ and the value of γ, for which this minimum is attained.

3.2. Prove formula (3.10).

3.3. Prove that conditions (3.8) and (2.14) are equivalent, as it was stated in the text.

3.4. Investigate convergence of iterations (3.1) for the equation

$$\frac{x}{x^2+1} = 0$$

depending on a choice of initial point x_0.

3.5. Consider the functional

$$\Phi(x) = \sum_{k=1}^{\infty} \frac{(x^{(k)})^2}{k^2}$$

in the Hilbert space

$$H_1 = l_2 = \left\{ x = (x^{(1)}, x^{(2)}, \ldots) : \sum_{k=1}^{\infty} (x^{(k)})^2 < \infty \right\},$$

with the scalar product

$$(x, y)_{l_2} = \sum_{k=1}^{\infty} x^{(k)} y^{(k)}.$$

The functional Φ has a unique minimizer in the space l_2, $x^* = (0, 0, \ldots)$, and $\Phi(x^*) = 0$. Show that Φ is *strictly convex*, i.e.,

$$\Phi(tx + (1-t)y) < t\Phi(x) + (1-t)\Phi(y), \quad x \ne y, \quad t \in (0, 1).$$

Construct a sequence $\{x_n\}_{n=0}^{\infty} \subset l_2$, such that

$$\lim_{n \to \infty} \Phi(x_n) = 0, \tag{3.23}$$

but $\|x_n\|_{l_2} = 1$, $n = 0, 1, \ldots$. Construct a sequence $\{x_n\}_{n=0}^{\infty} \subset l_2$ satisfying (3.23), such that $\|x_n\|_{l_2} \to \infty$ as $n \to \infty$.

3.6. Let $H_1 = \mathbb{R}^n$. Prove that, if the set

$$\{x \in \mathbb{R}^n : \Phi(x) \le \Phi(x_0)\}$$

is bounded, then the sequence $\{x_n\}$, defined by

$$x_{n+1} = x_n - \gamma \Phi'(x_n), \quad n = 0, 1, \ldots, \quad \gamma > 0,$$

has limiting points, and each limiting point $\bar{x}$ is a stationary point for Φ, i.e., $\Phi'(\bar{x}) = 0$.

3 The gradient method

3.7. Let an operator F take the form $F(x) = Ax - f$, where $A \in \mathcal{L}(H_1, H_2)$, $f \in H_2$. Write process (3.1) for this case. Prove that the functional Φ, corresponding to the above operator, is convex. What conditions on the operator A guarantee strict convexity of Φ? What conditions on A guarantee strong convexity of Φ? Recall, that a functional Φ is said to be *strongly convex*, if for any $x, y \in H_1$ there exists $\epsilon > 0$ such that

$$\Phi(tx + (1-t)y) \leq t\Phi(x) + (1-t)\Phi(y) - \epsilon \|x - y\|_{H_1}^2 \quad \text{for any } t \in (0, 1).$$

3.8. Write process (3.1) for the systems of equations from Problem 2.6. Is condition (3.8) satisfied for these systems?

3.9. Use formula (2.18) to write process (3.1) for equation (1.26). What can you say about the convergence of this process? Answer the same questions for the equation with operator (1.27).

3.10. Investigate whether or not it is possible to apply method (3.1) to the equations with operators (1.28) and (1.29).

4
Tikhonov's scheme

4.1 The Tikhonov functional

As we have seen above, the problem of solving equation (1.1) numerically is closely related to the problem of generating a convergent minimizing sequence for functional (2.1). Partly, classical iterative methods solve this problem in the regular case, i.e., one gets a convergent minimizing sequence, if the initial approximation point is chosen successfully. For example, gradient process (3.1) induces a sequence, which converges to a solution of equation (1.1) at a linear rate. If (1.1) is irregular, then it is not possible, in general, to obtain a convergent minimizing sequence using the algorithms, presented in Chapters 1–3.

Formally, the construction of a convergent minimizing sequence for arbitrary functional (2.1) was done by A. N. Tikhonov in the 1960s. Although numerical implementation of Tikhonov's scheme can be performed effectively for convex Φ only, modern iterative methods for solving irregular problems on the class $\mathcal{F}(N_1, N_2)$ exploit the basic ideas of Tikhonov's scheme considerably. In this chapter we give a brief account of Tikhonov's procedure, and investigate the issue of its practical realization.

Assume that equation (1.1) is solvable, and let $X^*(F)$ be the set of its solutions. Obviously, the set of minimizers of functional (2.1) coincides with $X^*(F)$. In addition to Φ, examine the functional

$$\Phi_\alpha(x) = \Phi(x) + \frac{\alpha}{2} \|x - \xi\|_{H_1}^2 = \frac{1}{2} \|F(x)\|_{H_2}^2 + \frac{\alpha}{2} \|x - \xi\|_{H_1}^2, \quad x \in H_1, \quad (4.1)$$

which is called the *Tikhonov functional* for equation (1.1). Here $\alpha > 0$ is a regularization parameter and ξ is some fixed element of the space H_1. It is clear, that

$$\inf_{x \in H_1} \Phi_\alpha(x) \geq 0 \quad \forall \alpha \geq 0.$$

The functional Φ_α does not necessarily have a global minimizer. However, for any $\alpha > 0$ and for any $\varepsilon > 0$ there exists an element $x_\alpha^\varepsilon \in H_1$, such that $\Phi_\alpha(x_\alpha^\varepsilon)$ differs from the infimum by not more than ε, i.e.,

$$\inf_{x \in H_1} \Phi_\alpha(x) \leq \Phi_\alpha(x_\alpha^\varepsilon) \leq \inf_{x \in H_1} \Phi_\alpha(x) + \varepsilon.$$

Suppose that $\varepsilon = \varepsilon(\alpha)$ depends on the regularization parameter α in such a way that

$$\lim_{\alpha \to 0} \frac{\varepsilon(\alpha)}{\alpha} = 0. \tag{4.2}$$

Take some $x^* \in X^*(F)$. Recall, that by our assumption, $X^*(F) \neq \emptyset$. By the definition of x_α^ε,

$$\Phi_\alpha(x_\alpha^\varepsilon) \leq \Phi_\alpha(x^*) + \varepsilon(\alpha).$$

Therefore, by (4.1) one gets

$$\Phi(x_\alpha^\varepsilon) + \frac{\alpha}{2} \|x_\alpha^\varepsilon - \xi\|_{H_1}^2 \leq \frac{\alpha}{2} \|x^* - \xi\|_{H_1}^2 + \varepsilon(\alpha). \tag{4.3}$$

From (4.3) it follows that

$$\Phi(x_\alpha^\varepsilon) \leq \frac{\alpha}{2} \|x^* - \xi\|_{H_1}^2 + \varepsilon(\alpha), \tag{4.4}$$

$$\|x_\alpha^\varepsilon - \xi\|_{H_1}^2 \leq \|x^* - \xi\|_{H_1}^2 + \frac{2\varepsilon(\alpha)}{\alpha}. \tag{4.5}$$

For the sake of simplicity, the set of elements $\{x_\alpha^{\varepsilon(\alpha)}\}_{\alpha \in (0, \alpha_0]}$ ($\alpha_0 > 0$) is called a sequence throughout this section. By (4.2)

$$\lim_{\alpha \to 0} \varepsilon(\alpha) = 0.$$

Hence, (4.4) implies that the sequence $x_\alpha^{\varepsilon(\alpha)}$ is *minimizing* for the functional Φ as $\alpha \to 0$, i.e.,

$$\lim_{\alpha \to 0} \Phi(x_\alpha^{\varepsilon(\alpha)}) = \inf_{x \in H_1} \Phi(x) = 0.$$

4.2 Properties of a minimizing sequence

1) First, let us prove that, under some additional assumption on the operator F, the sequence $x_\alpha^{\varepsilon(\alpha)}$ converges strongly to the set $X^*(F)$ in a sense that

$$\lim_{\alpha \to 0} \operatorname{dist}(x_\alpha^{\varepsilon(\alpha)}, X^*(F)) = 0,$$

where, by definition, for any $D \subset H_1$

$$\operatorname{dist}(x, D) = \inf_{y \in D} \|x - y\|_{H_1}, \quad x \in H_1.$$

Note, that by (4.5), the set of elements $\{x_\alpha^{\varepsilon(\alpha)}\}_{\alpha \in (0,\alpha_0]}$ is bounded in the norm of H_1 for any $\alpha_0 > 0$. Take arbitrary sequence of regularization parameters $\{\alpha_n\}$, satisfying the condition

$$\lim_{n\to\infty} \alpha_n = 0.$$

By virtue of its boundedness, the corresponding sequence $\{x_{\alpha_n}^{\varepsilon(\alpha_n)}\}$ is *weakly compact*, i.e., any infinite subset of its elements contains a subsequence that converges weakly in H_1 to some element of H_1. Thus, without loss of generality, one may conclude that the sequence $\{x_{\alpha_n}^{\varepsilon(\alpha_n)}\}$ itself converges weakly to an element $z^* \in H_1$. Suppose now that the operator F in (1.1) is *weakly continuous*, i.e., it maps any sequence $\{z_n\}$, weakly convergent in H_1 to some element $\bar{z}$, to the sequence $\{F(z_n)\}$, weakly convergent in H_2 to $F(\bar{z})$ (as $n \to \infty$). Unfortunately, the weak continuity of an operator does not follow, in general, from its *strong continuity*, i.e., the continuity of a map $F : H_1 \to H_2$ in the norms of H_1 and H_2. An example of a weakly continuous operator is given in Problem 4.15 below.

Under the assumption that F in equation (1.1) is weakly continuous, inequality (4.4) implies $z^* \in X^*(F)$. Indeed, since $\{x_{\alpha_n}^{\varepsilon(\alpha_n)}\}$ converges weakly in H_1 to z^*, the sequence $\{F(x_{\alpha_n}^{\varepsilon(\alpha_n)})\}$ converges weakly in H_2 to $F(z^*)$. The norm in a Hilbert space is weakly lower semi-continuous. Hence,

$$\|F(z^*)\|_{H_2} \leq \liminf_{n\to\infty} \|F(x_{\alpha_n}^{\varepsilon(\alpha_n)})\|_{H_2}.$$

By estimate (4.4),

$$\frac{1}{2}\|F(x_{\alpha_n}^{\varepsilon(\alpha_n)})\|_{H_2}^2 \leq \frac{\alpha_n}{2}\|x^* - \xi\|_{H_1}^2 + \varepsilon(\alpha_n),$$

and therefore

$$\lim_{n\to\infty} \|F(x_{\alpha_n}^{\varepsilon(\alpha_n)})\|_{H_2} = 0.$$

Thus, $F(z^*) = 0$ and $z^* \in X^*(F)$. Besides,

$$x_{\alpha_n}^{\varepsilon(\alpha_n)} - \xi \to z^* - \xi, \quad n \to \infty, \tag{4.6}$$

weakly in H_1 as $n \to \infty$. Let us show, that the sequence $\{x_{\alpha_n}^{\varepsilon(\alpha_n)} - \xi\}$, in fact, converges strongly to $z^* - \xi$. Inequalities (4.4) and (4.5) are satisfied for any element from the solution set $X^*(F)$, in particular, for the element z^*. So, (4.2) and (4.5) yield

$$\limsup_{n\to\infty} \|x_{\alpha_n}^{\varepsilon(\alpha_n)} - \xi\|_{H_1} \leq \|z^* - \xi\|_{H_1}. \tag{4.7}$$

On the other hand, from (4.6) it follows that

$$\|z^* - \xi\|_{H_1} \leq \liminf_{n\to\infty} \|x_{\alpha_n}^{\varepsilon(\alpha_n)} - \xi\|_{H_1}. \tag{4.8}$$

Applying (4.7) and (4.8), one derives

$$\lim_{n\to\infty} \|x_{\alpha_n}^{\varepsilon(\alpha_n)} - \xi\|_{H_1} = \|z^* - \xi\|_{H_1}. \tag{4.9}$$

Weak convergence (4.6) in H_1 and convergence (4.9) of the norms imply that $x_{\alpha_n}^{\varepsilon(\alpha_n)} - \xi$ converges strongly to $z^* - \xi$ as $n \to \infty$, and

$$\lim_{n\to\infty} \|x_{\alpha_n}^{\varepsilon(\alpha_n)} - z^*\|_{H_1} = 0.$$

This limit results in the identity

$$\lim_{\alpha\to 0} \inf_{x^* \in X^*(F)} \|x_\alpha^{\varepsilon(\alpha)} - x^*\|_{H_1} = 0, \tag{4.10}$$

which completes the proof.

2) Now let us verify, that with the operator F being weakly continuous, the functional Φ_α has a global minimizer, i.e., $\forall \alpha > 0$

$$\inf_{x \in H_1} \Phi_\alpha(x) \tag{4.11}$$

is attained at some element from H_1, and one can take $\varepsilon(\alpha) = 0$ in the argument above. In order to prove that, consider an arbitrary minimizing sequence $\{x_n\}$ for the functional Φ_α:

$$\lim_{n\to\infty} \Phi_\alpha(x_n) = d_\alpha := \inf_{x \in H_1} \Phi_\alpha(x).$$

For any $\omega > 0$, there exists $N(\omega)$ such, that for every $n \geq N(\omega)$ the following inequality holds

$$\Phi_\alpha(x_n) \leq d_\alpha + \omega.$$

Therefore, using (4.1), one gets the estimates, similar to (4.4) and (4.5):

$$\frac{1}{2}\|F(x_n)\|_{H_2}^2 \leq d_\alpha + \omega, \quad \|x_n - \xi\|_{H_1}^2 \leq \frac{2(d_\alpha + \omega)}{\alpha}, \quad n \geq N(\omega).$$

Hence, $\{x_n\}$ and $\{F(x_n)\}$ are bounded, and without loss of generality, one concludes that these sequences converge weakly in H_1 and H_2 to the elements $\tilde{x}$ and $\tilde{f}$, respectively. Weak continuity of the operator F in (1.1) implies that $\tilde{f} = F(\tilde{x})$, and one has

$$d_\alpha \leq \Phi_\alpha(\tilde{x}) = \frac{1}{2}\|F(\tilde{x})\|_{H_2}^2 + \frac{\alpha}{2}\|\tilde{x} - \xi\|_{H_1}^2$$

$$\leq \liminf_{n\to\infty} \left(\frac{1}{2}\|F(x_n)\|_{H_2}^2 + \frac{\alpha}{2}\|x_n - \xi\|_{H_1}^2\right) \leq d_\alpha + \omega.$$

Since $\omega > 0$ is arbitrary, this estimate yields $d_\alpha = \Phi_\alpha(\tilde{x})$, and $\inf_{x \in H_1} \Phi_\alpha(x)$ is attained at the element $\tilde{x}$.

3) The process we describe here, aimed at the construction of a minimizing sequence for the functional Φ, is called the Tikhonov scheme. Note, that for a generic operator $F \in \mathcal{F}(N_1, N_2)$, identity (4.10) cannot be replaced with a statement about either strong or weak convergence of the sequence $\{x_\alpha^{\varepsilon(\alpha)}\}$ as $\alpha \to 0$. In this sense, (4.10) is not possible to improve (see Problem 4.10). If the solution set $X^*(F)$ consists of a unique point x^*, then (4.10) implies that $\{x_\alpha^{\varepsilon(\alpha)}\}$ converges to x^*:

$$\lim_{\alpha \to 0} \|x_\alpha^{\varepsilon(\alpha)} - x^*\|_{H_1} = 0.$$

If there is more than one point in $X^*(F)$, then, for a weakly continuous F, identity (4.10) can be specified as follows. Denote by $X^{**}(F)$ the set of all elements in $X^*(F)$, nearest to ξ, i.e.,

$$X^{**}(F) = \{x^{**} \in X^*(F) : \|x^{**} - \xi\|_{H_1} = \text{dist}(\xi, X^*(F))\}.$$

Using (4.5), one can prove that $X^{**}(F) \neq \emptyset$ in the same way one proves the existence of a global minimizer for the functional Φ_α (see Problem 4.11). Then one arrives at the following refinement of (4.10):

$$\lim_{\alpha \to 0} \text{dist}(x_\alpha^{\varepsilon(\alpha)}, X^{**}(F)) = 0. \tag{4.12}$$

4.3 Other types of convergence

The properties of a minimizing sequence, presented in Section 4.2, have been obtained under the assumption that the operator F in (1.1) is weakly continuous. Here we would like to demonstrate a technique that allows one not to use this assumption.

Let us replace the weak continuity of F by the condition of a compact embedding of H_1 into some metric space (X, ρ_X). Namely, suppose H_1 is included into the space X element by element, i.e., each element of the initial space H_1 is also an element of X. The *compact embedding condition* amounts to the compactness of the operator $\text{Id} : H_1 \to X$, identifying an arbitrary element $x \in H_1$ with the same element, but viewed as an element of X. In other words, it is required that any subset of H_1, bounded in the norm of H_1, is compact as a subset of X with the metric ρ_X. One should keep in mind, that the limit of a sequence of elements from H_1, convergent in X, may not be an element of H_1.

Consider now our operator F as acting from X to H_2. This operator is naturally defined on all elements of X, that belong to H_1. Assume there is a continuous extension of F onto the space X. In practically interesting cases, this assumption is usually fulfilled. Define the extension by $\bar{F}$, and let $X^*(\bar{F})$ be the solution set for the equation $\bar{F}(x) = 0$ in the space X. In general, this set is bigger than the solution set $X^*(F)$ of the initial equation in the space H_1. Let us go back to $\{x_\alpha^{\varepsilon(\alpha)}\}$, and consider this sequence as a sequence of elements in X. By definition of the operator $\bar{F}$,

$$\bar{F}(x_\alpha^{\varepsilon(\alpha)}) = F(x_\alpha^{\varepsilon(\alpha)}), \quad \alpha > 0. \tag{4.13}$$

4 Tikhonov's scheme

According to (4.5), one can extract a subsequence from any sequence $\alpha_n \to 0$ ($n \to \infty$) in such a way, that the corresponding elements $x_{\alpha_n}^{\varepsilon(\alpha_n)}$ converge in X to some element $\bar{x} \in X$. By (4.13), continuity of $\bar{F}$, and inequality (4.4), $\bar{x}$ is contained in $X^*(\bar{F})$. Hence, the sequence $\{x_\alpha^{\varepsilon(\alpha)}\}$ converges in the metric of X to the set $X^*(\bar{F})$ as $\alpha \to 0$, i.e.,

$$\lim_{\alpha \to 0} \inf_{z \in X^*(\bar{F})} \rho_X(z, x_\alpha^{\varepsilon(\alpha)}) = 0. \tag{4.14}$$

This relation is similar to (4.10), but convergence of the approximations $x_\alpha^{\varepsilon(\alpha)}$ is established in the metric ρ_X, which generates a weaker topology as compared to the topology of the norm in the original space H_1. One can see, that for successful implementation of the above idea, the operator F must admit a continuous extension from H_1 onto a suitable metric space X.

Here is a typical example that illustrates how the proposed scheme can be applied in practice. Consider equation (1.1) with the following integral operator

$$[F(x)](t) = \int_a^b \varphi(t, s, x(s)) \, ds - f(t), \quad t \in (a, b), \tag{4.15}$$

where $\varphi = \varphi(t, s, \sigma)$ and $f = f(t)$ are some given functions. Take $W_2^1(a, b)$ as H_1, the Sobolev space of functions form $L_2(a, b)$ with square integrable generalized derivatives. Let $H_2 = L_2(a, b)$. It has been proven in a classical functional analysis course that the space $W_2^1(a, b)$ is compactly embedded in $C[a, b]$, the space of functions continuous on $[a, b]$. See, for example, [71]. If f and φ are continuous functions of their arguments, then operator (4.15) is defined on $W_2^1(a, b)$ as well as on $C[a, b]$, and it is continuous in both spaces as acting to $L_2(a, b)$. Therefore the required continuation $\bar{F}$ is described by the same formula (4.15), just like the initial operator F. Thus, Tikhonov's scheme, which guarantees the construction of a minimizing sequence $\{x_\alpha^{\varepsilon(\alpha)}\}$ ($\alpha \to 0$), weakly convergent in $W_2^1(a, b)$, also ensures strong convergence of this sequence in $C[a, b]$ to $X^*(\bar{F})$, the solution set for the equation $\bar{F}(x) = 0$ in the space $C[a, b]$.

In conclusion, we would like to point out, that the above method of getting a convergent minimizing sequence is not constructive, in most cases. The reason for that is the lack of general effective algorithms for computing the elements $x_\alpha^{\varepsilon(\alpha)}$, or $x_\alpha = x_\alpha^0$, for arbitrary operators $F \in \mathcal{F}(N_1, N_2)$. Already for $H_1 = H_2 = \mathbb{R}$, the Tikhonov functional Φ_α may have numerous local minima (see Problem 4.14), and the problem of finding $x_\alpha^{\varepsilon(\alpha)}$ ($\varepsilon \geq 0$) becomes a problem of global optimization. However, there is an important class of equations (1.1), for which the elements x_α can be calculated numerically in a rather effective manner. We shall consider this class in our next chapter.

4.4 Equations with noisy data

Let us briefly discuss the case of a perturbed operator F in equation (1.1). To this end, instead of (1.1), consider the following equation

$$F(x) = f, \quad x \in H_1 \quad (f \in H_2). \tag{4.16}$$

Suppose that f in the right-hand side of (4.16) is given by its δ-approximation $\tilde{f} \in H_2$, i.e.,

$$\|\tilde{f} - f\|_{H_2} \le \delta. \tag{4.17}$$

Let the error estimate δ be given, and our previous assumptions on the operator F be satisfied. In place of the exact discrepancy functional

$$\Phi(x) = \frac{1}{2}\|F(x) - f\|_{H_2}^2, \quad x \in H_1,$$

one now has

$$\tilde{\Phi}(x) = \frac{1}{2}\|F(x) - \tilde{f}\|_{H_2}^2, \quad x \in H_1.$$

The corresponding Tikhonov functional takes the form

$$\tilde{\Phi}_\alpha(x) = \tilde{\Phi}(x) + \frac{\alpha}{2}\|x - \xi\|_{H_1}^2 = \frac{1}{2}\|F(x) - \tilde{f}\|_{H_2}^2 + \frac{\alpha}{2}\|x - \xi\|_{H_1}^2, \quad x \in H_1.$$

In accordance with the above Tikhonov's scheme, fix $\alpha, \varepsilon > 0$ and define the elements x_α^ε from the condition

$$x_\alpha^\varepsilon \in H_1, \quad \inf_{x \in H_1} \tilde{\Phi}_\alpha(x) \le \tilde{\Phi}_\alpha(x_\alpha^\varepsilon) \le \inf_{x \in H_1} \tilde{\Phi}_\alpha(x) + \varepsilon. \tag{4.18}$$

Here we cannot expect x_α^ε, satisfying (4.18), to somehow approximate the solution set $X^*(F)$ of equation (4.16) as $\alpha, \varepsilon \to 0$. Since the discrepancy functional of the original equation is given with a fixed error, δ, the scalars α and ε must not only be in agreement with one another, but also with this error, and therefore their values should not be very small. Indeed, it does not make much of a sense to minimize the perturbed Tikhonov functional with the accuracy, substantially exceeding the error, with which this functional is known. Thus, we regard α and ε as functions of δ, i.e., $\alpha = \alpha(\delta), \varepsilon = \varepsilon(\delta)$. Following the logic of Section 4.1 and using inequality (4.17), instead of (4.4) and (4.5) one gets

$$\|F(x_\alpha^\varepsilon) - \tilde{f}\|_{H_2}^2 \le \delta^2 + \alpha\|x^* - \xi\|_{H_1}^2 + 2\varepsilon, \tag{4.19}$$

$$\|x_\alpha^\varepsilon - \xi\|_{H_1}^2 \le \|x^* - \xi\|_{H_1}^2 + \frac{2\varepsilon + \delta^2}{\alpha}, \tag{4.20}$$

where x^* is an arbitrary element of the set $X^*(F)$. It is clear, that estimates (4.19) and (4.20) can be obtained from (4.4) and (4.5), if one replaces ε with $\varepsilon + \delta^2/2$.

4 Tikhonov's scheme

This observation allows one to formulate the conditions on $\alpha = \alpha(\delta)$ and $\varepsilon = \varepsilon(\delta)$, sufficient for the convergence (as $\delta \to 0$) of $x_{\alpha(\delta)}^{\varepsilon(\delta)}$ to $X^{**}(F)$, the set of solutions to (4.16), nearest to the element ξ. Namely, let the operator F be weakly continuous. If

$$\lim_{\delta \to 0} \alpha(\delta) = \lim_{\delta \to 0} \frac{\delta^2}{\alpha(\delta)} = \lim_{\delta \to 0} \frac{\varepsilon(\delta)}{\alpha(\delta)} = 0, \qquad (4.21)$$

then

$$\lim_{\delta \to 0} \sup_{\tilde{f} : \|\tilde{f} - f\|_{H_2} \le \delta} \mathrm{dist}(x_{\alpha(\delta)}^{\varepsilon(\delta)}, X^{**}(F)) = 0, \qquad (4.22)$$

which is similar to (4.12).

Problems

4.1. Is F belonging to the class $\mathcal{F}(N_1, N_2)$ necessary for the validity of identities (4.10) and (4.14)?

4.2. Prove relations (4.10) and (4.14).

4.3. Show that, regardless of whether or not equation (1.1) is solvable,

$$\lim_{\alpha \to 0} \Phi(x_\alpha^{\varepsilon(\alpha)}) = d := \inf_{x \in H_1} \Phi(x).$$

Hint: Derive inequality, similar to (4.3), with x^* replaced by x_ω^* such, that $d \le \Phi(x_\omega^*) \le d + \omega$, $\omega > 0$. Then choose ω as a function of α.

4.4. Prove that the set $X^*(F)$ is closed in H_1, and, if the operator F is weakly continuous, $X^*(F)$ is also weakly closed, i.e., the weak limit of any subsequence of its elements is contained in $X^*(F)$.

4.5. Prove that the set $X^*(\bar{F})$ is closed in X.

4.6. Which operator equation becomes the necessary condition of minimum for the functional, defined by (4.1) and (4.15)? *Hint*: Use the result from Problem 3.10.

4.7. Can one replace the space $C[a, b]$ with $L_2(a, b)$ in the above example related to operator (4.15)?

4.8. An operator $F : H_1 \to H_2$ is said to be *weakly closed*, if for any sequence $\{x_n\}$, conditions

$$x_n \to \tilde{x} \quad \text{weakly in } H_1 \quad \text{and} \quad F(x_n) \to \tilde{f} \quad \text{weakly in } H_2 \quad \text{as } n \to \infty$$

imply

$$\tilde{f} = F(\tilde{x}).$$

Prove that any weakly continuous operator is also weakly closed.

4.9. Show that the results of Chapter 4 remain valid, if we assume that F is weakly closed instead of assuming that it is weakly continuous.

4.10. Consider the operator $F : [-2, 2] \to \mathbb{R}$, $F(x) = x^2 - 1$. Construct its continuation on $\mathbb{R}$, which is in the class $\mathcal{F}(N_1, N_2)$ and whose values outside the interval $[-2, 2]$ are at least 1. Apply Tikhonov's scheme with $\xi = 0$ to the continued operator and show, that for any sequence $\alpha_n \to 0$ $(n \to \infty)$ the elements $x_{\alpha_n}^{\varepsilon(\alpha_n)}$ may not have a limit, if $\varepsilon(\alpha_n) > 0, n = 0, 1, \ldots$.

4.11. Prove that $X^{**}(F) \neq \emptyset$. *Hint*: Use the result of Problem 4.4.

4.12. Prove equality (4.12).

4.13. Show that the operator $F : [-1, 1] \to \mathbb{R}$,

$$F(x) = \begin{cases} x^5 \sin(1/x), & x \in [-1, 1]\setminus\{0\} \\ 0, & x = 0 \end{cases}$$

can be continued on all of $\mathbb{R}$ in such a way, that the resulting operator is in the class $\mathcal{F}(N_1, N_2)$.

4.14. Set $\xi = 0$ in (4.1) and investigate the Tikhonov functional for the operator, obtained in Problem 4.13. Prove that the number of local minima of this functional grows unboundedly as $\alpha \to 0$. Verify that, for sufficiently large values of $\alpha > 0$, the functional becomes strongly convex and, therefore, it has only one (global) minimum.

4.15. Let $V : H_1 \to H$ be a linear compact operator, $G : H \to H_2$ be a continuous operator, H be a Hilbert space. Show that the operator $F : H_1 \to H_2$ such that

$$F(x) = G(Vx), \quad x \in H_1,$$

takes any weakly convergent sequence to a strongly convergent sequence and, therefore, it is weakly continuous. Assuming that the operator G is differentiable, find the derivative $F'(x)$. Prove that, if G is from the class $\mathcal{F}(N_1, N_2)$ as a mapping from H to H_2, then F is also from that class, probably with different constants N_1 and N_2.

4.16. Prove that condition (4.21) implies (4.22).

5

Tikhonov's scheme for linear equations

5.1 The main convergence result

In this chapter we consider an important special case of equation (1.1), when the operator F takes the form

$$F(x) = Ax - f, \quad A \in \mathcal{L}(H_1, H_2), \quad f \in H_2. \tag{5.1}$$

In other words, we are talking about a linear operator equation

$$Ax = f, \quad x \in H_1. \tag{5.2}$$

The operator F, defined by (5.1), belongs to the class $\mathcal{F}(N_1, 0)$, $N_1 \geq \|A\|_{\mathcal{L}(H_1, H_2)}$. In this case it is not difficult to prove the existence of a unique global minimizer, $x_\alpha \in H_1$, for the functional (4.1), which now takes the form

$$\Phi_\alpha(x) = \frac{1}{2}\|Ax - f\|_{H_2}^2 + \frac{\alpha}{2}\|x - \xi\|_{H_1}^2, \quad x \in H_1.$$

Indeed, consider the gradient of this functional:

$$\Phi_\alpha'(x) = (A^*Ax - A^*f) + \alpha(x - \xi), \quad x \in H_1.$$

Here A^*A is a self-adjoint operator, acting from H_1 to H_1, and

$$(A^*Ax, x)_{H_1} = \|Ax\|_{H_2}^2 \geq 0 \quad \forall x \in H_1,$$

which implies that the operator A^*A is nonnegative. Therefore

$$(\Phi_\alpha'(x) - \Phi_\alpha'(y), x - y)_{H_1} \geq \alpha\|x - y\|_{H_1}^2 \quad \forall x, y \in H_1 \quad (\alpha > 0).$$

Thus, one concludes that, whenever F is given by (5.1), functional (4.1) is strongly convex. In classical optimization theory it is established, that for a strongly convex

5.1 The main convergence result

functional there exists a unique point of minimum on any convex closed set. Therefore, functional (4.1) has a unique minimizer, x_α, on H_1, and x_α can be found from the equation

$$\Phi'_\alpha(x) = 0, \quad x \in H_1.$$

In detail, one can write this equation as follows

$$(A^*A + \alpha E_1)x = A^*f + \alpha\xi. \tag{5.3}$$

The operator $A^*A + \alpha E_1$ in (5.3) is continuously invertible for any $\alpha > 0$. Moreover, using a number of techniques, one can show that

$$\|(A^*A + \alpha E_1)^{-1}\|_{\mathcal{L}(H_1, H_1)} \leq \frac{1}{\alpha}, \quad \alpha > 0. \tag{5.4}$$

The element x_α admits an explicit representation

$$x_\alpha = \alpha(A^*A + \alpha E_1)^{-1}\xi + (A^*A + \alpha E_1)^{-1}A^*f$$
$$= [E_1 - (A^*A + \alpha E_1)^{-1}A^*A]\xi + (A^*A + \alpha E_1)^{-1}A^*f. \tag{5.5}$$

Let the solution set $X^*(F)$ of equation (5.2) be nonempty. It is easy to check, that $X^*(F)$ is an affine subspace in H_1, i.e., the set of the form

$$L_a = \{x \in H_1 : x = a + y, \ y \in L\},$$

where a is a fixed element from H_1, L is a linear subspace in H_1. So, for any $x^* \in X^*(F)$ one gets

$$X^*(F) = \{x \in H_1 : x = x^* + y, \ y \in N(A)\} = N(A) + x^*.$$

Here and below

$$N(A) = \{x \in H_1 : Ax = 0\}$$

is a null-space (kernel) of the operator $A \in \mathcal{L}(H_1, H_2)$. The linear subspace $N(A)$ and the affine subspace $X^*(F)$ are both closed subsets of H_1. As it follows from the theorem about a projection of a point onto a convex closed set in a Hilbert space, there is a unique element $x^*_\xi \in X^*(F)$, neatest to ξ. Besides, the operator F, defined by (5.1), is weakly continuous. It has been verified in Chapter 4 (see (4.12)), that any sequence $\{\bar{x}_n\}$ such, that

$$\lim_{n \to \infty} \|A\bar{x}_n - f\|_{H_2} = 0,$$

$$\limsup_{n \to \infty} \|\bar{x}_n - \xi\|_{H_1} \leq \|x^* - \xi\|_{H_1} \quad \forall x^* \in X^*(F),$$

contains a subsequence, converging strongly to the set

$$X^{**}(F) = \{x^{**} \in X^*(F) : \|x^{**} - \xi\|_{H_1} = \text{dist}(\xi, X^*(F))\}.$$

In our case, according to (4.4) and (4.5), as $\{\bar{x}_n\}$ one can take $\{x_{\alpha_n}\}$ with $\lim_{n \to \infty} \alpha_n = 0$. Since for operator (5.1) the set $X^{**}(F)$ consists of a single element x^*_ξ, $X^{**}(F) = \{x^*_\xi\}$, one arrives at the conclusion

$$\lim_{\alpha \to 0} \|x_\alpha - x^*_\xi\|_{H_1} = 0. \tag{5.6}$$

5.2 Elements of spectral theory

In order to simplify our presentation, suppose that the operator A is compact. Recall some facts from analysis (see, for example, [28]). A self-adjoint compact nonnegative operator A^*A has a finite or a countable set of orthonormal eigenvectors e_k with the corresponding eigenvalues λ_k ($A^*Ae_k = \lambda_k e_k$), $k = 1, 2, \ldots$, and one can assume

$$\lambda_1 \geq \lambda_2 \geq \cdots \geq \lambda_k \geq \cdots > 0, \tag{5.7}$$

where any chain of equalities in (5.7) is necessarily finite. If there is a countable set of distinct eigenvalues λ_k, then $\lim_{k \to \infty} \lambda_k = 0$. The spectrum of the operator A^*A takes the form

$$\sigma(A^*A) = \left(\bigcup_k \{\lambda_k\}\right) \cup \{0\},$$

and

$$\sigma(A^*A) \subset [0, \|A\|^2_{\mathcal{L}(H_1, H_2)}].$$

An arbitrary element $u \in H_1$ can be decomposed in terms of orthonormal system $\{e_k\}$:

$$u = \sum_k (u, e_k)_{H_1} e_k + P_{N(A^*A)} u. \tag{5.8}$$

In this expansion, $P_{N(A^*A)}$ is an orthogonal projection onto the null-space $N(A^*A)$, and the second part of the expansion is orthogonal to the first one. Note, $N(A^*A) = N(A)$. Let $\Psi(\lambda)$ be a scalar real-valued function, defined on the segment $[0, \|A\|^2_{\mathcal{L}(H_1, H_2)}]$ and, therefore, on the spectrum $\sigma(A^*A)$. Then we set up a function $\Psi(A^*A)$ of the operator A^*A as follows. Assign an element $u \in H_1$ to the domain $D(\Psi(A^*A))$ of the operator $\Psi(A^*A)$, if the series

$$\sum_k \Psi(\lambda_k)(u, e_k)_{H_1} e_k$$

converges in the norm of H_1, and for any $u \in D(\Psi(A^*A))$ define

$$\Psi(A^*A)u = \sum_k \Psi(\lambda_k)(u, e_k)_{H_1} e_k + \Psi(0) P_{N(A^*A)} u. \tag{5.9}$$

The operator $\Psi(A^*A) : D(\Psi(A^*A)) \subset H_1 \to H_1$, introduced in (5.9), is not necessarily bounded. The necessary and sufficient condition for the operator $\Psi(A^*A)$ to be bounded is

$$\sup_{\lambda \in \sigma(A^*A)} |\Psi(\lambda)| < \infty.$$

In that case, $D(\Psi(A^*A)) = H_1$, identity (5.9) is satisfied for any $u \in H_1$, and

$$\|\Psi(A^*A)\|_{\mathcal{L}(H_1,H_1)} = \sup_{\lambda \in \sigma(A^*A)} |\Psi(\lambda)|. \tag{5.10}$$

In addition to that, the operator $\Psi(A^*A)$ is self-adjoint. Clearly, if the set of distinct eigenvalues λ_k in (5.7) is finite, then $D(\Psi(A^*A)) = H_1$ for any function

$$\Psi : [0, \|A\|^2_{\mathcal{L}(H_1,H_2)}] \to \mathbb{R}.$$

Formulas (5.8)–(5.10) can naturally be generalized to the case of an arbitrary self-adjoint nonnegative compact operator $B \in \mathcal{L}(H_1, H_1)$. One just has to put B instead of A^*A whenever it is relevant. For example, definition (5.9) takes the form

$$\Psi(B)u = \sum_k \Psi(\lambda_k)(u, e_k)_{H_1} e_k + \Psi(0) P_{N(B)} u,$$

where $\{(\lambda_k, e_k)\}$ is now the system of positive eigenvalues and the corresponding orthonormal eigenvectors of the operator B. The equalities

$$D(\Psi(B)) = H_1, \quad \Psi(B) \in \mathcal{L}(H_1, H_1), \quad \|\Psi(B)\|_{\mathcal{L}(H_1,H_1)} = \sup_{\lambda \in \sigma(B)} |\Psi(\lambda)|$$

hold if and only if

$$\sup_{\lambda \in \sigma(B)} |\Psi(\lambda)| < \infty.$$

Also, $\Psi(B)^* = \Psi(B)$.

5.3 Minimizing sequences for linear equations

In this section we show that special form (5.1) of the operator F allows one to prove relation (5.6) without using the results of Chapter 4. As $\Psi(\lambda)$ in (5.9) consider

$$\Theta(\lambda, \alpha) = \frac{1}{\lambda + \alpha}, \quad \lambda \geq 0, \tag{5.11}$$

where α is a fixed positive parameter. One can verify that for $\Theta = \Theta(\lambda, \alpha)$ expression (5.9) defines a bounded linear operator $\Theta(A^*A, \alpha)$, and this operator is equal to $(A^*A + \alpha E_1)^{-1}$. Identity (5.5) can now be reformulated as follows

$$x_\alpha = [E_1 - \Theta(A^*A, \alpha) A^*A]\xi + \Theta(A^*A, \alpha) A^* f, \quad \alpha > 0. \tag{5.12}$$

Let us check that both terms in the right-hand side of (5.12) have limits in the norm of H_1 as $\alpha \to 0$; the first term approaches the projection of ξ onto the null-space $N(A^*A)$ of the operator A^*A, and the second one approaches the minimal norm solution of (5.2), i.e., the element $x_0^* \in X^*(F)$.

36   5 Tikhonov's scheme for linear equations

1) Consider the first term. By (5.9) and (5.11), it can be written in the form

$$\alpha(A^*A + \alpha E_1)^{-1}\xi = \sum_k \frac{\alpha}{\lambda_k + \alpha}(\xi, e_k)_{H_1} e_k + P_{N(A^*A)}\xi.$$

Prove that

$$\lim_{\alpha \to 0} \left\| \sum_k \frac{\alpha}{\lambda_k + \alpha}(\xi, e_k)_{H_1} e_k \right\|_{H_1} = 0. \tag{5.13}$$

Indeed, if the sum in (5.13) contains a finite number of terms, then (5.13) is obviously true. Suppose, there are infinitely many terms. Fix any $\alpha > 0$. The square of the norm in the left-hand side of (5.13) represents the following series

$$\sum_{k=1}^{\infty} \frac{\alpha^2}{(\lambda_k + \alpha)^2}(\xi, e_k)_{H_1}^2 = R(\alpha).$$

By Bessel's inequality, the series

$$\sum_{k=1}^{\infty} (\xi, e_k)_{H_1}^2$$

converges. Hence, for any $\varepsilon > 0$ fixed, there exists a number $N = N(\varepsilon)$ such, that

$$\sum_{k=N(\varepsilon)+1}^{\infty} (\xi, e_k)_{H_1}^2 \leq \frac{\varepsilon}{2}.$$

Therefore,

$$\sum_{k=N(\varepsilon)+1}^{\infty} \frac{\alpha^2}{(\lambda_k + \alpha)^2}(\xi, e_k)_{H_1}^2 \leq \sum_{k=N(\varepsilon)+1}^{\infty} (\xi, e_k)_{H_1}^2 \leq \frac{\varepsilon}{2}.$$

So, for any $\alpha > 0$,

$$R(\alpha) \leq \sum_{k=1}^{N(\varepsilon)} \frac{\alpha^2}{(\lambda_k + \alpha)^2}(\xi, e_k)_{H_1}^2 + \frac{\varepsilon}{2}.$$

In this inequality, the first term is bounded from above by the product

$$\frac{\alpha^2}{(\lambda_{N(\varepsilon)} + \alpha)^2} \sum_{k=1}^{N(\varepsilon)} (\xi, e_k)_{H_1}^2,$$

which does not exceed $\varepsilon/2$ for a sufficiently small $\alpha > 0$. Thus, (5.13) is proven.

2) Let us investigate the second term in (5.12). It takes the form

$$z_\alpha = \Theta(A^*A, \alpha)A^*Ax^*, \quad \alpha > 0, \tag{5.14}$$

where x^* is an arbitrary solution of equation (5.2). Recall, that the set $X^*(F)$ is not empty by our assumption. Since $Ax^* = f$ for any $x^* \in X^*(F)$, the sequence $\{z_\alpha\}$ remains the same for any choice of x^*. According to (5.8) and (5.9), for z_α, defined in (5.14), one gets

$$z_\alpha = \sum_k \frac{\lambda_k}{\lambda_k + \alpha}(x^*, e_k)_{H_1} e_k.$$

Now consider

$$w^* = \sum_k (x^*, e_k)_{H_1} e_k.$$

Since $x^* \in H_1$, the element w^* is well-defined and, by (5.8), is orthogonal to the subspace $N(A^*A)$. Besides, we point out, that w^* does not, in fact, depend on $x^* \in X^*(F)$, and the following relation is satisfied

$$\lim_{\alpha \to 0} \|z_\alpha - w^*\|_{H_1} = 0, \tag{5.15}$$

which is the consequence of (5.13), if one replaces ξ with x^* in that identity. Now (5.8) yields

$$w^* = (E_1 - P_{N(A^*A)})x^*,$$

therefore w^* is also a solution to (5.2). In addition to that, the orthogonal expansion holds

$$x^* = (E_1 - P_{N(A^*A)})x^* + P_{N(A^*A)}x^* = w^* + P_{N(A^*A)}x^*. \tag{5.16}$$

Equation (5.16) implies that

$$\|w^*\|_{H_1} \leq \|x^*\|_{H_1} \quad \forall x^* \in X^*(F),$$

i.e., w^* is a minimal norm element of the set $X^*(F)$. By (5.13) and (5.15), the limit (as $\alpha \to 0$) of the sequence x_α, introduced in (5.5), exists and it is equal to

$$w^* + P_{N(A^*A)}\xi = x_\xi^*.$$

The last equality can also be established by a direct verification of the necessary and sufficient condition of minimum in the problem of finding an element $x_\xi^* \in X^*(F)$, nearest to ξ. This condition is of the form

$$(\xi - x_\xi^*, x^* - x_\xi^*)_{H_1} = 0 \quad \forall x^* \in X^*(F). \tag{5.17}$$

5.4 A priori agreement between the regularization parameter and the error for equations with perturbed right-hand sides

Frequently, when we are trying to approximate a solution to equation (5.2) in the process of mathematical modeling, the pair (A, f) is not known exactly. Instead, some other pair, $(\tilde{A}, \tilde{f})$, is available, where $\tilde{A}$ is a linear operator from H_1 to H_2, close to A in some sense, and $\tilde{f} \in H_2$ is close to f. In order to obtain x_α in this case, naturally, one uses $(\tilde{A}, \tilde{f})$ in place of (A, f) in formula (5.12). However, if the approximations x_α get constructed that way, one obtains the elements, which do not satisfy (5.6), in general. To analyze the situation in details, one has to be more specific about the connection between $(\tilde{A}, \tilde{f})$ and (A, f). In this lecture course we restrict ourselves to the case when the noisy pair takes the form $(A, \tilde{f})$ and

$$\|\tilde{f} - f\|_{H_2} \leq \delta. \tag{5.18}$$

In other words, it is assumed that we know the exact operator $A = \tilde{A}$, the right-hand side f is given by its approximation $\tilde{f}$, and the upper bound for the error $\|\tilde{f} - f\|_{H_2}$ is available. As the result, for any α fixed, one calculates $x_\alpha(\tilde{f})$ instead of $x_\alpha = x_\alpha(f)$, and $x_\alpha(\tilde{f})$ is defined by (5.3), (5.5) and (5.12) if one replaces f with $\tilde{f}$ there. For α and δ fixed, it is not difficult to estimate the proximity of $x_\alpha(f)$ and $x_\alpha(\tilde{f})$ in the norm of H_1. Indeed, it follows from (5.12) and (5.18) that

$$\|x_\alpha(\tilde{f}) - x_\alpha(f)\|_{H_1} \leq \|\Theta(A^*A, \alpha)A^*\|_{\mathcal{L}(H_2, H_1)} \delta. \tag{5.19}$$

To evaluate the norm in the right-hand side of this inequality, apply formula (5.9). Since A^*v is orthogonal to the null-space of the operator A^*A for any $v \in H_2$, one has

$$\Theta(A^*A, \alpha)A^*v = \sum_k \frac{1}{\lambda_k + \alpha}(A^*v, e_k)_{H_1} e_k + \frac{1}{\alpha} P_{N(A^*A)} A^*v$$

$$= \sum_k \frac{1}{\lambda_k + \alpha}(A^*v, e_k)_{H_1} e_k = \sum_k \frac{1}{\lambda_k + \alpha}(v, Ae_k)_{H_2} e_k.$$

Recall that, for any operator $B \in \mathcal{L}(H_2, H_1)$,

$$H_1 = N(B^*) \oplus \overline{R(B)}, \tag{5.20}$$

where

$$R(B) = \{x \in H_1 : x = By, y \in H_2\}$$

is the image of the operator B, the over-bar stands for the closure in the norm of H_1. The orthogonality of A^*v to the null-space of A^*A, $N(A^*A) = N(A)$, follows from (5.20) for $B = A^*$. Therefore

$$\|\Theta(A^*A, \alpha)A^*v\|_{H_1}^2 = \sum_k \frac{1}{(\lambda_k + \alpha)^2}(v, Ae_k)_{H_2}^2$$

$$= \sum_k \frac{\lambda_k}{(\lambda_k + \alpha)^2}\left(v, \frac{1}{\sqrt{\lambda_k}} Ae_k\right)_{H_2}^2.$$

The elements

$$\frac{1}{\sqrt{\lambda_k}} Ae_k, \quad k = 1, 2, \ldots \tag{5.21}$$

form an orthonormal system in the space H_2. So, by Bessel's inequality one obtains

$$\sum_k \left(v, \frac{1}{\sqrt{\lambda_k}} Ae_k\right)^2_{H_2} \leq \|v\|^2_{H_2}.$$

Hence

$$\|\Theta(A^*A,\alpha)A^*v\|_{H_1} \leq \sup_k \frac{\sqrt{\lambda_k}}{\lambda_k + \alpha} \|v\|_{H_2}, \quad \alpha > 0,$$

and

$$\|\Theta(A^*A,\alpha)A^*\|_{\mathcal{L}(H_2,H_1)} \leq \sup_k \frac{\sqrt{\lambda_k}}{\lambda_k + \alpha}$$

$$\leq \sup_{\lambda \in [0,\infty)} \frac{\sqrt{\lambda}}{\lambda + \alpha} = \frac{1}{2\sqrt{\alpha}}, \quad \alpha > 0. \qquad (5.22)$$

Finally, from (5.19) and (5.22) one derives

$$\|x_\alpha(\tilde{f}) - x_\alpha(f)\|_{H_1} \leq \frac{\delta}{2\sqrt{\alpha}}. \qquad (5.23)$$

Inequality (5.23) illustrates that, for any α fixed, approximation (5.5) is stable with respect to the perturbations of f. However, the approximation error is increasing as α is getting smaller, and it goes to infinity as $\alpha \to 0$. Also, estimate (5.23) yields

$$\|x_\alpha(\tilde{f}) - x^*_\xi\|_{H_1} \leq \|x_\alpha(f) - x^*_\xi\|_{H_1} + \frac{\delta}{2\sqrt{\alpha}}. \qquad (5.24)$$

As $\alpha \to 0$, the first term in the right-hand side of this inequality approaches zero by (5.6), while the second term tends to infinity. Still, if there is a set of approximations $\tilde{f}$, for which the corresponding $\delta \to 0$, then choosing $\alpha = \alpha(\delta)$ for every δ in such a way that

$$\lim_{\delta \to 0} \alpha(\delta) = \lim_{\delta \to 0} \frac{\delta}{\sqrt{\alpha(\delta)}} = 0, \qquad (5.25)$$

and applying (5.24), one arrives at the following conclusion

$$\lim_{\delta \to 0} \sup_{\tilde{f}: \|\tilde{f} - f\|_{H_2} \leq \delta} \|x_{\alpha(\delta)}(\tilde{f}) - x^*_\xi\|_{H_1} = 0. \qquad (5.26)$$

One can take, for instance, $\alpha(\delta) = C\delta$, $C > 0$, as a function satisfying condition (5.25). Relation (5.26) indicates that, for the appropriate choice of $\alpha(\delta)$, approximations $x_\alpha(\tilde{f})$ have the property, which is typical for approximate solutions of regular operator equations, i.e., $x_\alpha(\tilde{f})$ approach the exact solution as $\delta \to 0$.

The result is obtained by means of a special connection between the regularization parameter α and the error δ. It is obvious, however, that agreement (5.25) is not practical if one is solving particular equation (5.2) with a fixed perturbed right-hand side $\tilde{f}$ and the corresponding error δ. The above example of a function $\alpha = \alpha(\delta)$ shows that, *theoretically*, any value of $\alpha(\delta)$ works for an individual equation with a fixed δ, while the approximation $x_{\alpha(\delta)}(\tilde{f})$ depends on the choice of $\alpha(\delta)$ considerably. Therefore, statements like (5.26), derived under assumptions like (5.25), are of theoretical importance, primarily. According to (5.25), the function $\alpha(\delta)$ is assigned a priori, i.e., before one calculates approximations (5.12), and the results of these calculations do not affect the value of $\alpha(\delta)$. This is called an a priori agreement between the regularization parameter and the error. The above disadvantage of a priori agreements can partly be eliminated, if one selects $\alpha(\delta)$ based on calculations of $x_\alpha(\tilde{f})$ with some set of test values of $\alpha > 0$ instead of setting the value of $\alpha(\delta)$ in advance. Such rules of choosing the regularization parameter, called a posteriori rules, are often more flexible, because they take into consideration the particular input data $\tilde{f}$ that is associated with a problem in question. Recall, that general a priori rule (5.25) applies to all approximate right-hand sides $\tilde{f}$, satisfying (5.18). Below we present one of the simplest a posteriori rules for choosing α.

5.5 The discrepancy principle

We start with a remark on formula (5.12). Show that

$$\Theta(A^*A, \alpha)A^* = (\alpha E_1 + A^*A)^{-1}A^* = A^*(\alpha E_2 + AA^*)^{-1}. \tag{5.27}$$

Indeed, let

$$u = (\alpha E_2 + AA^*)^{-1}v, \quad v \in H_2. \tag{5.28}$$

To prove (5.27) it is sufficient to verify that

$$(\alpha E_1 + A^*A)^{-1}A^*v = A^*u$$

or, equivalently,

$$A^*v = (\alpha E_1 + A^*A)A^*u.$$

Clearly, the last expression follows directly from (5.28). Identity (5.27) can be written in the form

$$\Theta(A^*A, \alpha)A^* = A^*\Theta(AA^*, \alpha), \quad \alpha > 0. \tag{5.29}$$

Using definition (5.9), one can see that property (5.29) holds not only for function (5.11), but also for any function $\Theta(\lambda, \alpha)$, continuous with respect to $\lambda \in [0, \|A\|^2_{\mathcal{L}(H_1, H_2)}]$ for any fixed $\alpha > 0$. In order to check that, one has to approximate $\Theta(\cdot, \alpha)$ on the interval $[0, \|A\|^2_{\mathcal{L}(H_1, H_2)}]$ by a polynomial $Q_n(\cdot, \alpha)$ and note that (5.29) is fulfilled for any function $\Theta(\lambda, \alpha)$, which depends on λ polynomially.

5.5 The discrepancy principle

Taking that into account, one can represent (5.12) for equation (5.2) with an approximate right-hand side $\tilde{f}$ as

$$x_\alpha(\tilde{f}) = [E_1 - A^*\Theta(AA^*, \alpha)A]\xi + A^*\Theta(AA^*, \alpha)\tilde{f}. \tag{5.30}$$

Hence,

$$Ax_\alpha(\tilde{f}) - \tilde{f} = [E_2 - AA^*\Theta(AA^*, \alpha)](A\xi - \tilde{f}). \tag{5.31}$$

Denote by $g(\alpha, \tilde{f})$ the norm of the discrepancy for equation (5.2) on the element $x_\alpha(\tilde{f})$:

$$g(\alpha, \tilde{f}) = \|Ax_\alpha(\tilde{f}) - \tilde{f}\|_{H_2}. \tag{5.32}$$

Set also

$$\Lambda(\mu, \alpha) = 1 - \mu\Theta(\mu, \alpha), \quad \mu \in [0, \|A\|^2_{\mathcal{L}(H_1, H_2)}], \quad \alpha > 0.$$

Then, according to (5.31),

$$Ax_\alpha(\tilde{f}) - \tilde{f} = \Lambda(AA^*, \alpha)(A\xi - \tilde{f}).$$

Let μ_k and φ_k ($k = 1, 2, \ldots$) be positive eigenvalues and the corresponding orthonormal eigenvectors of the compact operator $AA^* \in \mathcal{L}(H_2, H_2)$. It is easy to see, that the operator AA^* is nonnegative, and for this operator the formula, similar to (5.9), holds. Using this formula, one gets from (5.32)

$$g^2(\alpha, \tilde{f}) = \sum_k \Lambda^2(\mu_k, \alpha)(A\xi - \tilde{f}, \varphi_k)^2_{H_2} + \Lambda^2(0, \alpha)\|P_{N(AA^*)}(A\xi - \tilde{f})\|^2_{H_2}.$$

Here $A\xi - \tilde{f} = A(\xi - x^*) - (\tilde{f} - f)$ for any $x^* \in X^*(F)$. By (5.20), the element $A(\xi - x^*)$ is orthogonal to the null-space $N(AA^*) = N(A^*)$. Therefore

$$g^2(\alpha, \tilde{f}) = \sum_k \Lambda^2(\mu_k, \alpha)(A\xi - \tilde{f}, \varphi_k)^2_{H_2} + \Lambda^2(0, \alpha)\|P_{N(AA^*)}(\tilde{f} - f)\|^2_{H_2}.$$
$$\tag{5.33}$$

In the case under consideration, the function $\Theta(\lambda, \alpha)$ is given by equality (5.11). So,

$$\Lambda(\mu, \alpha) = \frac{\alpha}{\mu + \alpha}. \tag{5.34}$$

Since the series

$$\sum_k (A\xi - \tilde{f}, \varphi_k)^2_{H_2}$$

converges, from (5.33) and (5.34) one concludes,

$$\lim_{\alpha \to 0} g(\alpha, \tilde{f}) = \|P_{N(AA^*)}(\tilde{f} - f)\|_{H_2} \le \|\tilde{f} - f\|_{H_2} \le \delta. \tag{5.35}$$

5 Tikhonov's scheme for linear equations

Observe that $\Lambda(\mu, \alpha)$ is continuous and monotonically increasing with respect to α for every $\mu > 0$. Hence $g(\alpha, \tilde{f})$ is also continuous and monotonically increasing function of the regularization parameter $\alpha > 0$, and

$$\lim_{\alpha \to \infty} g(\alpha, \tilde{f}) = \|A\xi - \tilde{f}\|_{H_2}. \tag{5.36}$$

Suppose $\|A\xi - \tilde{f}\|_{H_2} > b\delta$, where $b > 1$ is some fixed constant, and describe the following a posteriori rule for choosing the regularization parameter α: given $\tilde{f}$ and δ, we find $\alpha = \alpha(\delta)$ from the equation

$$g(\alpha, \tilde{f}) = b\delta. \tag{5.37}$$

Properties of the function $g(\alpha, \tilde{f})$ imply that equation (5.37) has a unique solution $\alpha(\delta) > 0$. Let us investigate the behavior of the corresponding approximations $x_{\alpha(\delta)}(\tilde{f})$ as $\delta \to 0$. For any $x^* \in X^*(F)$ by definition of $x_\alpha(\tilde{f})$ one has

$$\|Ax_{\alpha(\delta)}(\tilde{f}) - \tilde{f}\|_{H_2}^2 + \alpha(\delta)\|x_{\alpha(\delta)}(\tilde{f}) - \xi\|_{H_1}^2 \leq \|Ax^* - \tilde{f}\|_{H_2}^2 + \alpha(\delta)\|x^* - \xi\|_{H_1}^2.$$

From the above, applying (5.18), (5.32) and (5.37), one derives

$$b^2\delta^2 + \alpha(\delta)\|x_{\alpha(\delta)}(\tilde{f}) - \xi\|_{H_1}^2 \leq \delta^2 + \alpha(\delta)\|x^* - \xi\|_{H_1}^2.$$

Thus,

$$\|x_{\alpha(\delta)}(\tilde{f}) - \xi\|_{H_1} \leq \|x^* - \xi\|_{H_1} \quad \forall x^* \in X^*(F). \tag{5.38}$$

Also, (5.18) and (5.37) yield

$$\|Ax_{\alpha(\delta)}(\tilde{f}) - f\|_{H_2} \leq \|Ax_{\alpha(\delta)}(\tilde{f}) - \tilde{f}\|_{H_2} + \|\tilde{f} - f\|_{H_2} \leq (b+1)\delta. \tag{5.39}$$

If one repeats the argument preceding (5.6), then by (5.38) and (5.39) one gets

$$\lim_{\delta \to 0} \sup_{\tilde{f}: \|\tilde{f} - f\|_{H_2} \leq \delta} \|x_{\alpha(\delta)}(\tilde{f}) - x_\xi^*\|_{H_1} = 0.$$

Hence, a posteriori rule (5.37) for choosing the regularization parameter guarantees convergence of the approximations $x_{\alpha(\delta)}(\tilde{f})$ as $\delta \to 0$ in the sense of identity (5.26). This rule is usually called the *discrepancy principle* as a consequence of representation (5.32) of the function $g(\alpha, \tilde{f})$.

To solve equation (5.37) one can use any of the existing methods for solving equations with one unknown. Since many of these methods require differentiability of the function, which defines the equation (see, for example, (1.23)), it is advisable to study the differential properties of the function $g(\cdot, \tilde{f})$. As it follows directly from (5.3), the mapping from $(0, \infty)$ to H_1, which takes α to $x_\alpha(\tilde{f})$, is continuously differentiable and the derivative $x'_\alpha(\tilde{f}) \in H_1$ can be found from the equation

$$A^* A x'_\alpha(\tilde{f}) + \alpha x'_\alpha(\tilde{f}) = \xi - x_\alpha(\tilde{f}). \tag{5.40}$$

From differentiation point of view, instead of (5.37) it is more convenient to deal with the equation $g^2(\alpha, \tilde{f}) = b^2\delta^2$. Using (5.40) it is easy to establish continuous differentiability of the function $g^2(\cdot, \tilde{f})$ and get an explicit formula for its derivative.

5.6 Approximation of a quasi-solution

In the above discussion it was assumed that the discrepancy functional

$$\Phi(x) = \frac{1}{2}\|Ax - f\|^2_{H_2}, \quad x \in H_1,$$

attains its infimum in H_1, and the value of this infimum is zero. Suppose now that the infimum is attained, but $\inf_{x \in H_1} \Phi(x) > 0$. In that case,

$$X^*(F) = \{x^* \in H_1 : \Phi(x^*) = \inf_{x \in H_1} \Phi(x)\}$$

is the set of quasi-solutions (solutions in the sense of least squares) to equation (5.2). By the convexity of the functional Φ,

$$X^*(F) = \{x \in H_1 : \Phi'(x) = 0\} = \{x \in H_1 : A^*Ax = A^*f\}.$$

Hence, in this section, instead of equation (5.2), we consider the following equation

$$A^*Ax = A^*f, \quad x \in H_1, \tag{5.41}$$

as the starting point of our analysis. For the sake of simplicity, let $B = A^*A$, $b = A^*f$. Then (5.41) is of the form $Bx = b$ with the operator B being non-negative, compact and self-adjoint. Formula (5.3) can be written as

$$(B + \alpha E_1)x_\alpha = b + \alpha\xi. \tag{5.42}$$

Taking into account the definition of a function of a compact self-adjoint operator, one can represent (5.42) in a way, similar to (5.12):

$$x_\alpha = [E_1 - \Theta(B,\alpha)B]\xi + \Theta(B,\alpha)b \tag{5.43}$$

with the function Θ, defined in (5.11). After making the necessary changes, one can repeat our previous argument and get the analogs of all the above results for scheme (5.43). In particular, one can check that (5.43) generates approximations x_α, converging to x^*_ξ, the quasi-solution of equation (5.2), nearest to the element ξ, as $\alpha \to 0$.

Problems

5.1. Prove inequality (5.4).

5.2. Prove the identity $X^*(F) = N(A) + x^*$, where x^* is an arbitrary element of $X^*(F)$.

5.3. Show that operator (5.1) is weakly continuous.

5.4. Show that $N(A^*A) = N(A)$.

5.5. Prove identity (5.10).

5.6. Show that representations (5.5) and (5.12) are equivalent.

5.7. Verify identity (5.17).

5.8. Prove that elements (5.21) form an orthonormal system.

5.9. Let $A : L_2(a, b) \to L_2(c, d)$ be an integral operator

$$(Ax)(t) = \int_a^b K(t, s)x(s)\, ds, \quad t \in (c, d),$$

and K be a continuous function on $[a, b] \times [c, d]$. Write equation (5.3) for that case.

5.10. Show that $\|A^*A\|_{\mathcal{L}(H_1, H_1)} = \|A\|^2_{\mathcal{L}(H_1, H_2)}$ for any operator $A \in \mathcal{L}(H_1, H_2)$.

5.11. Prove identity (5.36).

5.12. Derive (5.40).

5.13. Show that the mapping $\alpha \to x_\alpha(\tilde{f})$ ($\alpha > 0$) is twice differentiable and find its second derivative.

5.14. Prove that the function $h(\alpha, \tilde{f}) = g^2(\alpha, \tilde{f})$ is twice differentiable for $\alpha > 0$. Write Newton's method (1.23) for the equation $h(\alpha, \tilde{f}) = b^2\delta^2$.

5.15. Let B be a nonnegative self-adjoint compact operator in a Hilbert space H (this assumption holds for Problems 5.16 and 5.17 as well). Consider the functions $\Psi_1(\lambda) = 1$, $\Psi_2(\lambda) = \lambda$. Prove that $\Psi(B)$ is the identity operator E in the space H and $\Psi_2(B) = B$. Generalize the result to the case of $\Psi_3(\lambda) = \lambda^n$ for arbitrary $n \in \mathbb{N}$, i.e., show that $\Psi_3(B) = B^n$. Here $B^n = B \cdot B \cdots B$ (n times). Set $\Psi_4(\lambda) = (a + b\lambda)^n$, $a, b \in \mathbb{R}$, $n \in \mathbb{N}$. Prove that $\Psi_4(B) = (aE + bB)^n$.

5.16. Let the function $\Theta(\lambda, \alpha)$ be defined by identity (5.11). Prove that $\Theta(B, \alpha) = (B + \alpha E)^{-1}$, $\alpha > 0$.

5.17. Assume that $\Psi(\lambda) = e^{-t\lambda}$ ($t \geq 0$) and $x_0 \in H$. Define $e^{-tB} = \Psi(B)$. Verify that $e^{-tB} \in \mathcal{L}(H, H)$. Prove that the function $x(t) = e^{-tB}x_0$, which takes values in H, is a solution to the initial value problem $x'(t) + Bx(t) = 0$, $t \geq 0$, $x(0) = x_0$.

5.18. Prove that for any $f \in H_2$, i.e., regardless of solvability of equations (5.2) and (5.41),

$$\lim_{\alpha \to 0} \|A^*Ax_\alpha(f) - A^*f\|_{H_1} = 0.$$

5.19. Prove that the operator $(A^*A)^\nu$, $A \in \mathcal{L}(H_1, H_2)$, is nonnegative for any $\nu > 0$.

6
The gradient scheme for linear equations

6.1 The technique of spectral analysis

The technique of spectral analysis for linear compact self-adjoint operators, used in the previous chapter, makes it possible to derive some new important properties of the gradient method, which are different from those considered in Chapter 3. Gradient method (3.1) applied to equation (5.2) takes the form

$$x_{n+1} = x_n - \gamma(A^*Ax_n - A^*f), \quad n = 0, 1, \ldots \quad (\gamma > 0), \qquad (6.1)$$

where

$$x_0 = \xi \in H_1$$

is an initial approximation. In case of linear equations, estimates (1.2) and (1.3) are exact (unimprovable) for $N_1 = \|A\|_{\mathcal{L}(H_1, H_2)}$ and $N_2 = 0$. Following (3.15) and (3.17), let us assume that the step size γ in (6.1) satisfies the condition

$$0 < \gamma < \frac{2}{\|A\|^2_{\mathcal{L}(H_1, H_2)}}. \qquad (6.2)$$

Since

$$\sigma(A^*A) \subset [0, \|A\|^2_{\mathcal{L}(H_1, H_2)}],$$

condition (6.2) implies that

$$|1 - \gamma\lambda| < 1 \quad \forall \lambda \in \sigma(A^*A) \setminus \{0\} = \{\lambda_1, \lambda_2, \ldots\}. \qquad (6.3)$$

Proceeding by induction in n, one easily gets an explicit representation for x_n in terms of ξ and f:

$$x_n = (E_1 - \gamma A^*A)^n \xi + \gamma \sum_{k=0}^{n-1} (E_1 - \gamma A^*A)^k A^* f. \qquad (6.4)$$

6 The gradient scheme for linear equations

One can verify by direct calculations that

$$\gamma \sum_{k=0}^{n-1}(1-\gamma\lambda)^k = \frac{1-(1-\gamma\lambda)^n}{\lambda}, \quad \lambda \neq 0. \tag{6.5}$$

If $\lambda = 0$, then the value of the sum in (6.5) is $n\gamma$. It is appropriate therefore to introduce the function

$$\Theta(\lambda,\alpha) = \begin{cases} \lambda^{-1}\left(1-(1-\gamma\lambda)^{1/\alpha}\right), & \lambda \neq 0, \\ \alpha^{-1}\gamma, & \lambda = 0, \end{cases} \tag{6.6}$$

which is equal to the sum in the left-hand side of (6.5) with $\alpha = 1/n$, $n \in \mathbb{N}$, for any $\lambda \geq 0$. Clearly, the function $\Theta(\lambda,\alpha)$ depends on the spectral variable λ polynomially. Using notation (6.6) and the result of Problem 5.15, one can represent identity (6.4) in the form, analogous to (5.12),

$$x_n = [E_1 - \Theta(A^*A,\alpha)A^*A]\xi + \Theta(A^*A,\alpha)A^*f, \quad \alpha = 1/n, \quad n \in \mathbb{N}. \tag{6.7}$$

Thus, initial approximation $x_0 = \xi$ plays the same part in iterative scheme (6.1) as the element ξ in Tikhonov scheme (4.1), (5.3). In both cases this element is an estimate of the unknown solution x^*, and it is assigned based on some a priori information about x^*. In the absence of such information, one can simply set $\xi = 0$.

To investigate convergence of method (6.7), we apply the scheme, used in our previous chapter. Denote the first term in (6.7) by y_n and suppose

$$\Lambda(\lambda,\alpha) = 1 - \Theta(\lambda,\alpha)\lambda = (1-\gamma\lambda)^{1/\alpha}, \quad \lambda \geq 0, \quad \alpha > 0.$$

Then

$$y_n = \Lambda(A^*A,\alpha)\xi, \quad \alpha = 1/n,$$

and, according to (5.9),

$$y_n = \sum_k \Lambda(\lambda_k,\alpha)(\xi,e_k)_{H_1}e_k + P_{N(A^*A)}\xi. \tag{6.8}$$

As before, by $\{e_k\}$ we denote the family of orthonormal eigenvectors of the compact operator A^*A, corresponding to the eigenvalues

$$\lambda_1 \geq \lambda_2 \geq \cdots \geq \lambda_k \geq \cdots > 0.$$

In expansion (6.8), it is taken into consideration that $\Lambda(0,\alpha) = 1$. The square of the norm of the first term in sum (6.8) equals

$$\left\|\sum_k (1-\gamma\lambda_k)^{1/\alpha}(\xi,e_k)_{H_1}e_k\right\|_{H_1}^2 = \sum_k (1-\gamma\lambda_k)^{2/\alpha}(\xi,e_k)_{H_1}^2. \tag{6.9}$$

6.1 The technique of spectral analysis

If the sums in (6.9) contain a finite number of terms, then, by (6.3), both expressions in (6.9) converge to 0 as $\alpha \to 0$ or, equivalently, as $n = 1/\alpha \to \infty$. If the number of terms is infinite, then one estimates

$$R(\alpha) = \sum_{k=1}^{\infty}(1-\gamma\lambda_k)^{2/\alpha}(\xi, e_k)_{H_1}^2,$$

following the same scheme as in Chapter 5, and obtains

$$\lim_{\alpha \to 0} R(\alpha) = 0. \tag{6.10}$$

Hence,

$$\lim_{n\to\infty} \|y_n - P_{N(A^*A)}\xi\|_{H_1} = 0. \tag{6.11}$$

Now we introduce

$$z_n = \Theta(A^*A, \alpha)A^*f = \Theta(A^*A, \alpha)A^*Ax^*, \quad x^* \in X^*(F).$$

Definitions (5.9) and (6.6) yield

$$z_n = \sum_k (1-(1-\gamma\lambda_k)^{1/\alpha})(x^*, e_k)_{H_1} e_k.$$

Like in Chapter 5, introduce the notation

$$w^* = \sum_k (x^*, e_k)_{H_1} e_k.$$

It is already known that w^* is a minimal-norm element on the solution set $X^*(F)$ of equation (5.2), and it does not depend on our choice of $x^* \in X^*(F)$ For the square of the norm $\|z_n - w^*\|_{H_1}$ one can write the representation, similar to (6.9),

$$\|z_n - w^*\|_{H_1}^2 = \left\|\sum_k (1-\gamma\lambda_k)^{1/\alpha}(x^*, e_k)_{H_1} e_k\right\|_{H_1}^2 = \sum_k (1-\gamma\lambda_k)^{2/\alpha}(x^*, e_k)_{H_1}^2.$$

Applying the previous argument with ξ being replaced by x^*, one concludes that

$$\lim_{n\to\infty} \|z_n - w^*\|_{H_1} = 0. \tag{6.12}$$

Combining (6.11) and (6.12) as well as using the identity $x_n = y_n + z_n$, one gets

$$\lim_{n\to\infty} \|x_n - x_\xi^*\|_{H_1} = 0, \quad x_\xi^* = w^* + P_{N(A^*A)}\xi. \tag{6.13}$$

In Chapter 5, a similar relationship was justified for the approximations x_α, generated by Tikhonov's scheme. Thus, both methods give rise to the sequences, which converge to the point x_ξ^*, a solution of equation (5.2), nearest to the element ξ.

6.2 A priori stopping rule

Now we consider the behavior of iterations x_n when the right-hand side of equation (5.2) is given by its approximation $\tilde{f}$ satisfying (5.18). In this case, it is natural to use the following scheme for solving equation (5.2)

$$x_{n+1} = x_n - \gamma(A^*Ax_n - A^*\tilde{f}), \quad n = 0, 1, \ldots, \quad x_0 = \xi. \qquad (6.14)$$

One can apply formula (6.7) to the nth iterative point, defined by (6.14), provided that f is replaced with $\tilde{f}$ in (6.14). Denote this point by $x_n(\tilde{f})$. For the sake of consistency, by $x_n(f)$ we denote the element in original expression (6.7). Then (5.18) and (6.7) imply

$$\|x_n(\tilde{f}) - x_n(f)\|_{H_1} \leq \|\Theta(A^*A, \alpha)A^*\|_{\mathcal{L}(H_2, H_1)}\delta, \quad \alpha = n^{-1}. \qquad (6.15)$$

The above result is identical to (5.19), although $\Theta(\lambda, \alpha)$ is different. For the norm in the right-hand side of (6.15), one can easily get the estimate similar to (5.22). Like in the previous chapter, one obtains

$$\|\Theta(A^*A, \alpha)A^*\|_{\mathcal{L}(H_2, H_1)} \leq \sup_k \Theta(\lambda_k, \alpha)\sqrt{\lambda_k}$$

$$\leq \max_{\lambda \in [0, \|A\|^2_{\mathcal{L}(H_1, H_2)}]} \frac{1 - (1 - \gamma\lambda)^{1/\alpha}}{\sqrt{\lambda}}.$$

Using a number of arguments, one can verify that, under assumption (6.2), there exists a constant $C_1 > 0$, independent on α, such that

$$\max_{\lambda \in [0, \|A\|^2_{\mathcal{L}(H_1, H_2)}]} \frac{1 - (1 - \gamma\lambda)^{1/\alpha}}{\sqrt{\lambda}} \leq \frac{C_1}{\sqrt{\alpha}} = C_1\sqrt{n}, \quad \alpha = n^{-1}, \quad n \in \mathbb{N}. \qquad (6.16)$$

Combining (6.15) and (6.16), one derives

$$\|x_n(\tilde{f}) - x_n(f)\|_{H_1} \leq C_1\sqrt{n}\delta. \qquad (6.17)$$

The constant C_1 in (6.17) does not depend on n and δ. Estimate (6.17) yields

$$\|x_n(\tilde{f}) - x_\xi^*\|_{H_1} \leq \|x_n(f) - x_\xi^*\|_{H_1} + C_1\sqrt{n}\delta. \qquad (6.18)$$

If one chooses a number $n = n(\delta)$ such that

$$\lim_{\delta \to 0} n(\delta) = \infty, \quad \lim_{\delta \to 0} \sqrt{n(\delta)}\,\delta = 0, \qquad (6.19)$$

then, by (6.13) and (6.19), one has

$$\lim_{\delta \to 0} \sup_{\tilde{f}: \|\tilde{f} - f\|_{H_2} \leq \delta} \|x_{n(\delta)}(\tilde{f}) - x_\xi^*\|_{H_1} = 0. \qquad (6.20)$$

One can take, for example, $n(\delta) = [C\delta^{-1}]$ to satisfy (6.19), where C is an arbitrary positive constant and $[p]$ rounds the number p to the nearest integer towards zero. Equality (6.20) is a complete analogue of (5.26) related to Tikhonov's method. The number $n(\delta)$ in (6.19) provides the stopping rule for iterations (6.14) in case of noisy data. As it follows from (6.18), for a fixed $\delta > 0$, one cannot ensure the convergence of $x_n(\tilde{f})$ to a solution of (5.2) as $n \to \infty$. The same applies to Tikhonov's method, where the regularization parameter α plays the part of the iteration number n. In connection with this remark, we point out to the complete similarity of the second terms in the right-hand sides of estimates (5.24) and (6.18).

6.3 A posteriori stopping rule

Just like (5.25), a priori condition (6.19), which defines the stopping rule, is of asymptotic nature. Therefore in practice, it is often replaced with some a posteriori rules. Let us describe one of the most simple a posteriori rules, suitable for approximations (6.14). Take any $b > 1$. As in Chapter 5, assume $\|A\xi - \tilde{f}\|_{H_2} > b\delta$. Continue iterations (6.14) until the inequality

$$\|Ax_n(\tilde{f}) - \tilde{f}\|_{H_2} \leq b\delta \qquad (6.21)$$

is satisfied for the first time. Denote the number n, for which (6.21) holds for the first time, by $n(\delta)$. Since for $n = 0$ ($x_0 = \xi$) condition (6.21) is not fulfilled, the number $n(\delta)$, if exists, is different from zero, i.e., $n(\delta) \in \mathbb{N}$. Take the point $x_{n(\delta)}(\tilde{f})$ as an approximate solution to (5.2). Below we prove that, firstly, the number $n(\delta)$ always exists and, secondly, the constructed approximation $x_{n(\delta)}(\tilde{f})$ satisfies (6.20). As one can see from Problem 6.8, it makes sense to choose $b = 2$.

We start with some auxiliary results. It follows directly from (6.7) that, for any solution $x^* \in X^*(F)$ of equation (5.2), one has

$$x_n(\tilde{f}) - x^* = [E_1 - \Theta(A^*A, \alpha)A^*A](\xi - x^*) + \Theta(A^*A, \alpha)A^*(\tilde{f} - f). \qquad (6.22)$$

Using identity (5.29), which holds for function (6.6) as well, one gets

$$Ax_n(\tilde{f}) - \tilde{f} = A[E_1 - \Theta(A^*A, \alpha)A^*A](\xi - x^*) \\ - [E_2 - \Theta(AA^*, \alpha)AA^*](\tilde{f} - f). \qquad (6.23)$$

Let us point out to the following algebraic property of function (6.6):

$$\sup_{\lambda \in [0, \|A\|^2_{\mathcal{L}(H_1, H_2)}]} \lambda^p |1 - \Theta(\lambda, \alpha)\lambda| \leq C_2 \alpha^p \quad \forall \alpha > 0, \quad p > 0. \qquad (6.24)$$

Here the constant $C_2 = C_2(p)$ does not depend on α. Show that

$$\lim_{\alpha \to 0} \frac{1}{\sqrt{\alpha}} \|A[E_1 - \Theta(A^*A, \alpha)A^*A](\xi - x_\xi^*)\|_{H_2} = 0, \quad \alpha = n^{-1}, \quad n \in \mathbb{N}. \qquad (6.25)$$

6 The gradient scheme for linear equations

Transforming the square of the norm, which appears under the limit in (6.25), one obtains the equality

$$\|A[E_1 - \Theta(A^*A, \alpha)A^*A](\xi - x_\xi^*)\|_{H_2}^2$$
$$= (A[E_1 - \Theta(A^*A, \alpha)A^*A](\xi - x_\xi^*), A[E_1 - \Theta(A^*A, \alpha)A^*A](\xi - x_\xi^*))_{H_2}$$
$$= (\xi - x_\xi^*, \Psi(A^*A, \alpha)(\xi - x_\xi^*))_{H_1}, \qquad (6.26)$$

where

$$\Psi(\lambda, \alpha) = \lambda(1 - \Theta(\lambda, \alpha)\lambda)^2 = \lambda(1 - \gamma\lambda)^{2/\alpha}. \qquad (6.27)$$

Definition (5.9) implies that

$$(\xi - x_\xi^*, \Psi(A^*A, \alpha)(\xi - x_\xi^*))_{H_1} = \sum_k \Psi(\lambda_k, \alpha)(\xi - x_\xi^*, e_k)_{H_1}^2. \qquad (6.28)$$

If the sum in (6.28) contains a finite number of terms, then required property (6.25) holds, since, by (6.3),

$$\Psi(\lambda_k, \alpha) \le \lambda_1 q^{2/\alpha}, \quad q \in (0, 1)$$

for every k. Consider the case of infinitely many terms. By the definition of x_ξ^*, the element $\xi - x_\xi^*$ is orthogonal to the null-space $N(A^*A)$. Setting $B = A^*A$ in (5.20), one verifies that $\xi - x_\xi^* \in \overline{R(A^*A)}$. Therefore, for any $\varepsilon > 0$ there exist elements $u = u(\varepsilon)$, $v = v(\varepsilon) \in H_1$ such that

$$\xi - x_\xi^* = A^*Au + v, \quad \|v\|_{H_1} \le \varepsilon. \qquad (6.29)$$

Identities (6.28) and (6.29), combined with the elementary inequality

$$(a + b)^2 \le 2(a^2 + b^2), \qquad (6.30)$$

yield

$$(\xi - x_\xi^*, \Psi(A^*A, \alpha)(\xi - x_\xi^*))_{H_1}$$
$$\le 2 \sum_{k=1}^{\infty} \Psi(\lambda_k, \alpha)(A^*Au, e_k)_{H_1}^2 + 2 \sum_{k=1}^{\infty} \Psi(\lambda_k, \alpha)(v, e_k)_{H_1}^2. \qquad (6.31)$$

Since

$$(A^*Au, e_k)_{H_1} = (u, A^*Ae_k)_{H_1} = \lambda_k(u, e_k)_{H_1},$$

the first sum in (6.31) can be written in the form

$$\sum_{k=1}^{\infty} \lambda_k^2 \Psi(\lambda_k, \alpha)(u, e_k)_{H_1}^2.$$

According to (6.24) and (6.27),

$$\sup_{k\in\mathbb{N}} \lambda_k^2 \Psi(\lambda_k,\alpha) \le \Big(\sup_{\lambda\in[0,\|A\|^2_{\mathcal{L}(H_1,H_2)}]} \lambda^{3/2}|1-\Theta(\lambda,\alpha)\lambda| \Big)^2 \le C_3\alpha^3. \quad (6.32)$$

By Bessel's inequality

$$\sum_{k=1}^{\infty}(u,e_k)^2_{H_1} < \infty.$$

From (6.32) it follows that the first term in the right-hand side of inequality (6.31) can be estimated by $C_4\alpha^3$ with a suitable constant $C_4 = C_4(\varepsilon)$, which does not depend on $\alpha = n^{-1}$. Turning to the second sum in (6.31) and applying (6.24) once again, we note that

$$\sup_{k\in\mathbb{N}} \Psi(\lambda_k,\alpha) \le \Big(\sup_{\lambda\in[0,\|A\|^2_{\mathcal{L}(H_1,H_2)}]} \sqrt{\lambda}|1-\Theta(\lambda,\alpha)\lambda| \Big)^2 \le C_5\alpha.$$

Here the constant C_5 is independent on α and ε. Using Bessel's inequality one more time, we get

$$\sum_{k=1}^{\infty}(v,e_k)^2_{H_1} \le \|v\|^2_{H_1} \le \varepsilon^2.$$

Thus, the second term in (6.31) does not exceed $C_5\alpha\varepsilon^2$. Ultimately, we conclude that the value under the limit in (6.25) is bounded from above by $\sqrt{C_4(\varepsilon)}\alpha + \sqrt{C_5}\varepsilon$ and, as one passes to the limit with $\alpha \to 0$, one obtains $\sqrt{C_5}\varepsilon$. Since $\varepsilon > 0$ is arbitrary, this means that identity (6.25) is satisfied.

It is also important to point out to the estimate below, which follows directly from (5.10).

$$\|E_2 - \Theta(AA^*,\alpha)AA^*\|_{\mathcal{L}(H_2,H_2)} \le \sup_{\lambda\in[0,\|A\|^2_{\mathcal{L}(H_1,H_2)}]} |1-\Theta(\lambda,\alpha)\lambda| = 1. \quad (6.33)$$

The above auxiliary result allows us to justify a posteriori stopping rule (6.14). By (5.18), (6.23), (6.25) and (6.33), one has

$$\|Ax_n(\tilde{f}) - \tilde{f}\|_{H_2} \le C_6\sqrt{\alpha} + \delta, \quad (6.34)$$

where $\alpha = n^{-1}$, and the constant C_6 does not depend on α and δ. Since $b > 1$, for a sufficiently large $n \in \mathbb{N}$, the right-hand side of (6.34) is smaller than $b\delta$ and inequality (6.21) is fulfilled. From that point of view, rule (6.21) correctly defines some number $n = n(\delta) \in \mathbb{N}$. Note, that the following inequalities hold

$$\|A[E_1 - \Theta(A^*A,\alpha(\delta))A^*A](\xi - x_\xi^*)\|_{H_2} \le (b+1)\delta,$$
$$(b-1)\delta \le \|A[E_1 - \Theta(A^*A,\tilde{\alpha}(\delta))A^*A](\xi - x_\xi^*)\|_{H_2}. \quad (6.35)$$

6 The gradient scheme for linear equations

Here

$$\alpha(\delta) = n(\delta)^{-1} \quad \text{and} \quad \tilde{\alpha}(\delta) = \tilde{n}(\delta)^{-1}, \quad \tilde{n}(\delta) = n(\delta) - 1.$$

The first inequality in (6.35) is the consequence of (5.18), (6.21), (6.23) and (6.33). The second inequality follows from relation (6.23), if one applies this relation with the number $n = \tilde{n}(\delta) = n(\delta) - 1$ and the corresponding value of $\alpha = \tilde{\alpha}(\delta) = \tilde{n}(\delta)^{-1} > \alpha(\delta)$, and then uses (6.33) and the inequality

$$\|Ax_{\tilde{n}(\delta)}(\tilde{f}) - \tilde{f}\|_{H_2} > b\delta.$$

Recall, that the number $\tilde{n}(\delta) \geq 0$ is well-defined, because $n(\delta) \in \mathbb{N}$.

Without loss of generality one can assume that there are only two alternatives for $n(\delta)$ as $\delta \to 0$, i.e., the number $n(\delta) \to \infty$ as $\delta \to 0$, or this number remains bounded as δ approaches zero. In the first case one has

$$\lim_{\delta \to 0} \alpha(\delta) = \lim_{\delta \to 0} n(\delta)^{-1} = 0.$$

Hence, from (6.25) and the second inequality in (6.35) one derives

$$\lim_{\delta \to 0} \frac{\delta}{\sqrt{\tilde{\alpha}(\delta)}} = 0.$$

Since

$$\frac{\delta}{\sqrt{\alpha(\delta)}} \leq \frac{\sqrt{2}\delta}{\sqrt{\tilde{\alpha}(\delta)}} \quad \text{for } n(\delta) \geq 2,$$

one concludes

$$\lim_{\delta \to 0} \frac{\delta}{\sqrt{\alpha(\delta)}} = 0. \tag{6.36}$$

According to (6.15), (6.16) and (6.22),

$$\|x_{n(\delta)}(\tilde{f}) - x^*\|_{H_1} \leq \|[E_1 - \Theta(A^*A, \alpha(\delta))A^*A](\xi - x_\xi^*)\|_{H_1} + C_1 \frac{\delta}{\sqrt{\alpha(\delta)}}. \tag{6.37}$$

Together with (6.13) and (6.36), estimate (6.37) implies that $\|x_{n(\delta)}(\tilde{f}) - x^*\|_{H_1}$ goes to zero as $\delta \to 0$. If the number $n(\delta)$ does not approach infinity, then the value $\alpha(\delta) \geq \alpha_0 > 0$ for any $\delta > 0$ sufficiently small. Therefore the second term in (6.37) still tends to zero as $\delta \to 0$. On the other hand, the first inequality in (6.35) yields

$$\lim_{\delta \to 0} \|A[E_1 - \Theta(A^*A, \alpha(\delta))A^*A](\xi - x_\xi^*)\|_{H_2} = 0. \tag{6.38}$$

Keeping in mind that the element $\xi - x_\xi^*$ is orthogonal to the null-space $N(A^*A) = N(A)$, one can easily show that the last identity results in

$$\lim_{\delta \to 0} \|[E_1 - \Theta(A^*A, \alpha(\delta))A^*A](\xi - x_\xi^*)\|_{H_1} = 0. \tag{6.39}$$

Thus, by (6.37), $\|x_{n(\delta)}(\tilde{f}) - x^*\|_{H_1} \to 0$ as $\delta \to 0$. Summarizing the above, one concludes that a posteriori rule (6.21) generates approximations $x_{n(\delta)}(\tilde{f})$, which have property (6.20).

Problems

6.1. Prove identity (6.4).

6.2. Show that (6.10) holds.

6.3. Verify inequality (6.16).

6.4. Show that estimate (6.16) is exact, i.e., there is a constant $C > 0$ that does not depend on n such that

$$\max_{\lambda \in [0, \|A\|^2_{\mathcal{L}(H_1, H_2)}]} \frac{1 - (1 - \gamma\lambda)^n}{\sqrt{\lambda}} \geq C\sqrt{n} \quad \forall n \in \mathbb{N}.$$

6.5. Derive identities (6.22) and (6.23).

6.6. Prove estimate (6.24).

6.7. Show that (6.38) implies (6.39) if $\alpha(\delta) \geq \alpha_0 > 0$.

6.8. Let the condition $0 < \gamma \leq 1/\|A\|^2_{\mathcal{L}(H_1, H_2)}$, which is stronger than (6.2), be satisfied for scheme (6.14). Prove that if $\|Ax_n(\tilde{f}) - \tilde{f}\|_{H_2} > 2\delta$, then the following inequality is fulfilled

$$\|x_{n+1}(\tilde{f}) - x^*\|_{H_1} < \|x_n(\tilde{f}) - x^*\|_{H_1} \quad \forall x^* \in X^*(F).$$

Suppose $\dim X^*(F) > 0$. What can you say about the location of the points $x_n(\tilde{f})$ and $x_n(f)$ ($\delta = 0$) with respect to the affine subspace $X^*(F)$?

6.9. Obtain analogs of the results from the current chapter for the case when

$$\inf_{x \in H_1} \Phi(x) > 0, \quad X^*(F) = \{x^* \in H_1 : \Phi(x^*) = \inf_{x \in H_1} \Phi(x)\} \neq \emptyset.$$

How should one modify (6.16)–(6.19)?

6.10. Prove that for any element $f \in H_2$ the following identity holds

$$\lim_{n \to \infty} \|A^*Ax_n(f) - A^*f\|_{H_1} = 0.$$

7

Convergence rates for the approximation methods in the case of linear irregular equations

7.1 The source-type condition (STC)

Approximations (5.5) and (6.1) converge to a solution of equation (5.2). However, the convergence rate in relations (5.6) and (6.13) can be arbitrarily slow, in general. As the result, when the level of noise δ is fixed, the approximation errors for the elements $x_{\alpha(\delta)}(\tilde{f})$ and $x_{n(\delta)}(\tilde{f})$ remain unknown. Indeed, even though the first terms in the right-hand sides of estimates (5.24) and (6.18) tend to zero, the rate of this convergence is not possible to find. Therefore, expressions (5.26) and (6.20) are of asymptotic nature and do not provide any information about the errors $\|x_{\alpha(\delta)}(\tilde{f}) - x_\xi^*\|_{H_1}$ or $\|x_{n(\delta)}(\tilde{f}) - x_\xi^*\|_{H_1}$ for a specific problem with fixed $\tilde{f}$ and δ. This situation is typical when it comes to the approximation methods for irregular equations, see [13].

The lack of an estimate for the convergence rate is not so unusual in numerical analysis and approximation theory. A complete orthonormal system decomposition (for instance, a trigonometric system decomposition) of some function in L_2 is another classical example of convergence that can be arbitrarily slow (in L_2 norm). This is no longer the case, if it is additionally known that the function under consideration has some degree of smoothness. For trigonometric decomposition, smoothness means that the function we are trying to approximate is n times differentiable. Under the smoothness assumption, the convergence rate of the decomposition can be effectively estimated. For approximations (5.5) and (6.1) the situation is similar. If the object of the approximation, which is a solution x_ξ^* to equation (5.2) in our case, has certain additional properties, it turns out that one can not only estimate the rates of convergence of x_α and x_n to x_ξ^* with respect to α and n, but also the errors of approximation $\|x_{\alpha(\delta)}(\tilde{f}) - x_\xi^*\|_{H_1}$ or $\|x_{n(\delta)}(\tilde{f}) - x_\xi^*\|_{H_1}$ with respect to δ.

We start by considering approximation (5.5). Let x_ξ^* be a solution to equation (5.2), nearest to some fixed element ξ. From (5.5) it follows that

$$\begin{aligned} x_\alpha - x_\xi^* &= (A^*A + \alpha E_1)^{-1}[A^*Ax_\xi^* + \alpha\xi - (A^*A + \alpha E_1)x_\xi^*] \\ &= \alpha(A^*A + \alpha E_1)^{-1}(\xi - x_\xi^*), \quad \alpha > 0. \end{aligned} \qquad (7.1)$$

Clearly, if one replaces x_ξ^* with any solution $x^* \in X^*(F)$, identity (7.1) is still fulfilled. Since the element $\xi - x_\xi^*$ is orthogonal to the null-space $N(A)$, by (5.20) one has

$$\xi - x_\xi^* \in \overline{R(A^*)}. \tag{7.2}$$

Suppose a stronger condition holds

$$\xi - x_\xi^* \in R(A^*),$$

i.e., the elements ξ and x_ξ^* satisfy the identity

$$\xi - x_\xi^* = A^* v, \tag{7.3}$$

where v is some element of the space H_2. Substituting (7.3) into (7.1) and using estimate (5.22), one gets

$$\|x_\alpha - x_\xi^*\|_{H_1} \leq \frac{\sqrt{\alpha}}{2} \|v\|_{H_2}. \tag{7.4}$$

Inequality (7.4) together with (5.23) imply

$$\|x_\alpha(\tilde{f}) - x_\xi^*\|_{H_1} \leq \frac{\sqrt{\alpha}}{2} \|v\|_{H_2} + \frac{\delta}{2\sqrt{\alpha}}. \tag{7.5}$$

The expression in the right-hand side of (7.5) attains its minimum with respect to $\alpha > 0$ when

$$\alpha = \alpha(\delta) = \|v\|_{H_2}^{-1} \delta.$$

For this value of $\alpha = \alpha(\delta)$, the right-hand side of (7.5) is equal to $\sqrt{\|v\|_{H_2} \delta}$. Thus, one can specify (5.26) as follows

$$\sup_{\tilde{f}:\|\tilde{f}-f\|_{H_2}\leq\delta} \|x_{\alpha(\delta)}(\tilde{f}) - x_\xi^*\|_{H_1} \leq \sqrt{\|v\|_{H_2} \delta}. \tag{7.6}$$

Estimate (7.6) allows one to see the quality of the approximation of x_ξ^* by the element $x_{\alpha(\delta)}(\tilde{f})$, obtained for a specific problem with fixed $\tilde{f}$ and δ.

Let us discuss the above condition (7.3). For a particular solution x_ξ^*, a priori fixed, identity (7.3) indicates that the element ξ must be contained in a non-closed affine subspace $R(A^*)_{x_\xi^*}$ in H_1, passing though the point x_ξ^*. This condition can be relaxed. Namely, it is enough to assume that the following identity is satisfied

$$(E_1 - P_{N(A^*A)})(\xi - x^*) = A^* \bar{v}, \quad \bar{v} \in H_2, \tag{7.7}$$

for some solution $x^* \in X^*(F)$. In this case, applying (7.1) and (5.9), one has

$$x_\alpha - x^* = \alpha(A^*A + \alpha E_1)^{-1} A^* \bar{v} + \alpha(A^*A + \alpha E_1)^{-1} P_{N(A^*A)}(\xi - x^*)$$
$$= \alpha(A^*A + \alpha E_1)^{-1} A^* \bar{v} + P_{N(A^*A)}(\xi - x^*). \tag{7.8}$$

It follows from (7.8) that the approximation x_α fails to converge to x^* as $\alpha \to 0$ and, instead, it converges to another solution

$$x^* + P_{N(A^*A)}(\xi - x^*) = \bar{x}_\xi^* \in X^*(F).$$

Estimates (7.4)–(7.6), with x_ξ^* being replaced by $\bar{x}_\xi^*$, remain valid. Formula (7.7) is equivalent to the condition

$$(E_1 - P_{N(A^*A)})(\xi - x^*) \in R(A^*). \tag{7.9}$$

Note that if one modifies (7.9) by putting $\overline{R(A^*)}$ instead of $R(A^*)$, then one gets a weaker assumption, which is fulfilled for every ξ and x^*. This is a consequence of orthogonal expansion (5.20). Hence, condition (7.9) holds for 'almost all' (in the sense of density) elements $\xi \in H_1$. A similar remark applies to (7.3) that enhances (from the same point of view) inclusion (7.2), which is always true. Obviously, for a particular ξ the norm $\|v\|_{H_2}$ can be very large, and this will have a negative impact on the quality of estimates (7.4)–(7.6). Therefore, assuming that a solution x_ξ^* is fixed, it is worthwhile to provide an upper bound d on $\|v\|_{H_2}$. Then identity (7.3), along with the condition $\|v\|_{H_2} \leq d$, serve as requirements on the choice of ξ. According to these requirements, ξ must belong to the ellipsoid

$$S_d = \{x \in H_1 : x = x_\xi^* + A^*v, \ \|v\|_{H_2} \leq d\}, \tag{7.10}$$

centered at the point x_ξ^*. It is important to keep in mind, that in case of a compact operator A, acting from an infinite dimentional space H_1, ellipsoid (7.10), which is a compact subset of H_1, cannot contain any ball of a positive radius. Moreover, the center of this ellipsoid x_ξ^* is unknown. Thus, one can see that in practice a problem of finding an element form S_d is not much simpler than original equation (5.2). The above difficulty can formally be removed if the element ξ is somehow a priori fixed. Then condition (7.3) (or (7.9)) turns into an a priori assumption on the solution x_ξ^* (or x^*). Suppose ξ is chosen to be zero. In this case (7.3) is reduced to the inclusion $x_0^* \in R(A^*)$. This is called a *source-type condition* on $x^* = x_0^*$. For an arbitrary ξ, connected with x_ξ^* by identity (7.3), one has a *source-type condition* on the initial discrepancy $\xi - x_\xi^*$.

If we assume that source-type condition (7.3) is satisfied together with the condition $\|v\|_{H_2} \leq d$, i.e., $\xi \in S_d$, then it follows immediately from (7.5) and (7.6) that if

$$\alpha = \alpha(\delta) = d^{-1}\delta,$$

then

$$\sup_{\tilde{f}:\|\tilde{f}-f\|_{H_2} \leq \delta} \|x_{\alpha(\delta)}(\tilde{f}) - x_\xi^*\| \leq \sqrt{d\delta}.$$

Here, as we establish the dependence between α and δ, we use some upper bound d on the norm of v, which is specified a priori, instead of using the norm of v itself, since in most cases $\|v\|$ is unknown in practice. The same level of accuracy

$\|x_{\alpha(\delta)}(\tilde{f}) - x_\xi^*\|_{H_1}$ as $\delta \to 0$ is attained for any choice of $\alpha = \alpha(\delta)$ such that

$$C_1 d^{-1}\delta \le \alpha(\delta) \le C_2 d^{-1}\delta,$$

where the constants $0 < C_1 \le C_2$ do not depend on δ.

Note, that it is the irregularity of equation (5.2) with a compact operator A, which results in the necessity to introduce condition (7.3). Indeed, suppose the equation is regular, i.e., the operator A, or the operator A^*A, is continuously invertible. In both cases, the solution set $X^*(F)$ consists of a unique point $x^* = A^{-1}f$, or $x^* = (A^*A)^{-1}A^*f$, and the spectrum of the operator A^*A is separated from zero

$$\inf_{\lambda \in \sigma(A^*A)} \lambda = \varepsilon > 0. \tag{7.11}$$

Hence, instead of (5.4) one has the estimate

$$\|(A^*A + \alpha E_1)^{-1}\|_{\mathcal{L}(H_1,H_1)} \le \frac{1}{\alpha + \varepsilon} \le \frac{1}{\varepsilon} \quad \forall \alpha \ge 0. \tag{7.12}$$

In place of (5.22) the following estimate holds

$$\|(A^*A + \alpha E_1)^{-1}A^*\|_{\mathcal{L}(H_2,H_1)} \le \frac{\|A^*\|_{\mathcal{L}(H_2,H_1)}}{\alpha + \varepsilon} \le \frac{\|A^*\|_{\mathcal{L}(H_2,H_1)}}{\varepsilon} \quad \forall \alpha \ge 0.$$

Therefore, (7.1) and (5.19) yield

$$\|x_\alpha(f) - x^*\|_{H_1} \le C_3\alpha, \quad \|x_\alpha(\tilde{f}) - x_\alpha(f)\|_{H_1} \le C_4\delta.$$

By choosing the value of $\alpha(\delta)$ proportional to δ, $\alpha(\delta) = C_5\delta$, one obtains

$$\|x_\alpha(\tilde{f}) - x^*\|_{H_1} \le C_6\delta. \tag{7.13}$$

The last estimates are satisfied without any source-type conditions. Inequality (7.13) looks natural, since, if calculations are reasonable, the approximation error is of the same order as the error in the input data. This property is typical for regular problems only. It is significant, that estimate (7.6), derived in irregular case under the source-type assumption on the initial discrepancy, is worse than (7.13) considering the degree of δ.

7.2 STC for the gradient method

Let us investigate the convergence rate for iterations (6.1). Suppose that for the initial element ξ and the corresponding solution x_ξ^* to equation (5.2) a priori condition (7.3) holds. Then from (6.7) one gets

$$x_n - x_\xi^* = [E_1 - \Theta(A^*A, \alpha)A^*A](\xi - x_\xi^*) = [E_1 - \Theta(A^*A, \alpha)A^*A]A^*v.$$

7 Convergence rates in the case of linear irregular equations

Recall that the function $\Theta(\lambda,\alpha)$ is defined in (6.6), $\alpha = n^{-1}, n \in \mathbb{N}$. From the above one obtains

$$\|x_n - x_\xi^*\|_{H_1} \leq \|[E_1 - \Theta(A^*A,\alpha)A^*A]A^*\|_{\mathcal{L}(H_2,H_1)}\|v\|_{H_2}. \tag{7.14}$$

Applying identities (2.11) and

$$\Theta(A^*A,\alpha)^* = \Theta(A^*A,\alpha)$$

(see Chapter 5), write

$$\|[E_1 - \Theta(A^*A,\alpha)A^*A]A^*\|_{\mathcal{L}(H_2,H_1)} = \|A[E_1 - \Theta(A^*A,\alpha)A^*A]\|_{\mathcal{L}(H_1,H_2)}.$$

Using the same argument as in (6.26), one concludes that for any element $u \in H_1$

$$\|A[E_1 - \Theta(A^*A,\alpha)A^*A]u\|_{H_2}^2 = (u, \Psi(A^*A,\alpha)u)_{H_1}$$
$$\leq \|\Psi(A^*A,\alpha)\|_{\mathcal{L}(H_1,H_1)}\|u\|_{H_1}^2,$$

where the function $\Psi(\lambda,\alpha)$ is given by (6.27). By (5.10) and (6.24) one has

$$\|\Psi(A^*A,\alpha)\|_{\mathcal{L}(H_1,H_1)} \leq \sup_{\lambda \in [0,\|A\|_{\mathcal{L}(H_1,H_2)}^2]} \lambda|1 - \Theta(\lambda,\alpha)\lambda|^2 \leq C_7\alpha, \quad \alpha = n^{-1}.$$

Finally, one derives

$$\|[E_1 - \Theta(A^*A,\alpha)A^*A]A^*\|_{\mathcal{L}(H_2,H_1)} \leq C_8\sqrt{\alpha} = \frac{C_8}{\sqrt{n}},$$

$$\|x_n - x_\xi^*\|_{H_1} \leq \frac{C_8}{\sqrt{n}}\|v\|_{H_2}, \quad n \in \mathbb{N}. \tag{7.15}$$

Thus, by (7.15) and (6.17),

$$\|x_n(\tilde{f}) - x_\xi^*\|_{H_1} \leq C_9\left(\frac{1}{\sqrt{n}}\|v\|_{H_2} + \sqrt{n}\delta\right). \tag{7.16}$$

This estimate is similar to (7.5). The expression in the right-hand side of (7.16) attains its minimum with respect to $n \in \mathbb{N}$ at the point $n = n(\delta)$, where $n(\delta)$ is equal to $[\delta^{-1}\|v\|_{H_2}]$, or $[\delta^{-1}\|v\|_{H_2}] + 1$. The corresponding minimal value of the right-hand side in (7.16) is of order $O(\sqrt{\|v\|_{H_2}\delta})$. Hence, with the above $n(\delta)$, one obtains

$$\sup_{\tilde{f}:\|\tilde{f}-f\|_{H_2}\leq\delta} \|x_{n(\delta)}(\tilde{f}) - x_\xi^*\|_{H_1} \leq C_{10}\sqrt{\|v\|_{H_2}\delta}.$$

If $\xi \in S_d$, then using the stopping rule $n(\delta)$ such that

$$C_{11}[d\delta^{-1}] \leq n(\delta) \leq C_{12}[d\delta^{-1}] \quad (0 < C_{11} \leq C_{12}),$$

one gets

$$\sup_{\tilde{f}:\|\tilde{f}-f\|_{H_2}\leq\delta} \|x_{n(\delta)}(\tilde{f}) - x_\xi^*\|_{H_1} \leq C_{13}\sqrt{d\delta}.$$

7.3 The saturation phenomena

Are there any other conditions, different form (7.3), that guarantee some convergence rates for the approximations in question? The answer is yes. Consider a one parametric family of the source-type conditions in the form

$$\xi - x_\xi^* = (A^*A)^\nu u, \quad u \in H_1. \tag{7.17}$$

The linear continuous self-adjoint operator $(A^*A)^\nu$ ($\nu > 0$) is defined by formula (5.9) with $\Psi(\lambda) = \lambda^\nu$:

$$(A^*A)^\nu u = \sum_k \lambda_k^\nu (u, e_k)_{H_1} e_k, \quad u \in H_1. \tag{7.18}$$

Since the elements e_k are orthogonal to the null-space $N(A^*A) = N(A)$, it follows from (5.20) and (7.18) that

$$(A^*A)^\nu u \in \overline{R(A^*)}, \quad \nu > 0.$$

Therefore, source-type condition (7.17) is an assumption on the solution x_ξ^*, enhancing the requirement on x_ξ^* to be the nearest to ξ (see(7.2)). One can understand the constant ν in (7.17) as the level of 'smoothness' of the initial discrepancy $\xi - x_\xi^*$. If, as before, we fix the solution x_ξ^* and require that u in (7.17) is bounded by some constant d, then (7.17) is reduced to the condition on ξ to be contained in the ellipsoid

$$\Sigma_d = \{x \in H_1 : x = x_\xi^* + (A^*A)^\nu u, \|u\|_{H_1} \leq d\}. \tag{7.19}$$

The above argument about the difficulties involved in finding an element from S_d in practice can also be applied to the ellipsoid (7.19).

One can show that for $\nu = 1/2$, assumption (7.17) is equivalent to (7.3) in the sense that

$$R[(A^*A)^{1/2}] = R(A^*).$$

Hence one can view (7.17) as some generalization of (7.3).

Let us study the behavior of approximations (5.5) under condition (7.17). Instead of estimate (7.4) one has

$$\|x_\alpha - x_\xi^*\|_{H_1} \leq \max_{\lambda \in [0, \|A\|^2_{\mathcal{L}(H_1, H_2)}]} \frac{\alpha \lambda^\nu}{\lambda + \alpha} \|u\|_{H_1}. \tag{7.20}$$

By direct calculations one can see that

$$\max_{\lambda \in [0, \|A\|^2_{\mathcal{L}(H_1, H_2)}]} \frac{\alpha \lambda^\nu}{\lambda + \alpha} = \begin{cases} O(\alpha^\nu), & 0 < \nu \leq 1, \\ O(\alpha), & \nu \geq 1. \end{cases} \tag{7.21}$$

Thus for $\nu \geq 1$, the right-hand side in estimate (7.20) is $O(\alpha)$ regardless of the value of ν. This means that, for any $\nu > 0$, the order of $\|x_\alpha - x_\xi^*\|_{H_1}$, guaranteed by (7.20) and (7.21), does not exceed $O(\alpha)$. Estimate (7.5) is changed to

$$\|x_\alpha(\tilde{f}) - x_\xi^*\|_{H_1} \leq C_{14}\|u\|_{H_1}\alpha^\nu + \frac{\delta}{2\sqrt{\alpha}} \qquad (7.22)$$

for $0 < \nu \leq 1$, and to

$$\|x_\alpha(\tilde{f}) - x_\xi^*\|_{H_1} \leq C_{15}\|u\|_{H_1}\alpha + \frac{\delta}{2\sqrt{\alpha}} \qquad (7.23)$$

for $\nu \geq 1$. For δ fixed and ν given, it is worthwhile to minimize the right-hand side of estimates (7.22) or (7.23) with respect to $\alpha > 0$. With such an optimal choice of $\alpha = \alpha(\delta)$ the resulting error estimate $\|x_{\alpha(\delta)}(\tilde{f}) - x_\xi^*\|_{H_1}$ takes the form

$$\sup_{\tilde{f}:\|\tilde{f}-f\|_{H_2}\leq\delta} \|x_\alpha(\tilde{f}) - x_\xi^*\|_{H_1} = \begin{cases} O(\|u\|_{H_1}^{1/(2\nu+1)}\delta^{2\nu/(2\nu+1)}), & 0 < \nu \leq 1 \\ O(\|u\|_{H_1}^{1/3}\delta^{2/3}), & \nu \geq 1. \end{cases} \qquad (7.24)$$

Assuming that $\xi \in \Sigma_d$, it is not difficult to formulate a more practical rule of choosing α as a function of δ, which uses the value of d only and provides the error estimate of the same order.

It is remarkable that, no matter how one chooses α as a function of δ for approximations (5.5), the estimate $O(\delta^{2/3})$ in the right-hand side of (7.24) is the best possible on the whole class $\mathcal{F}(N_1, 0)$. On smaller subclasses of $\mathcal{F}(N_1, 0)$, one can get the estimate $O(\delta)$ that is typical for regular problems (see (7.13)).

When it comes to approximations (6.1), the situation is slightly different. Namely, under condition (7.17) with any $\nu > 0$ fixed, one gets the estimate similar to (7.14)

$$\|x_n - x_\xi^*\|_{H_1} \leq \|[E_1 - \Theta(A^*A, \alpha)A^*A](A^*A)^\nu\|_{\mathcal{L}(H_1, H_1)}\|u\|_{H_1}.$$

Here, according to (6.24),

$$\|[E_1 - \Theta(A^*A, \alpha)A^*A](A^*A)^\nu\|_{\mathcal{L}(H_1, H_1)} \leq \sup_{\lambda \in [0, \|A\|_{\mathcal{L}(H_1, H_2)}^2]} \lambda^\nu|1 - \Theta(\lambda, \alpha)\lambda| \leq C_{16}\alpha^\nu.$$

Therefore,

$$\|x_n - x_\xi^*\|_{H_1} \leq C_{16}\|u\|_{H_1}\alpha^\nu = \frac{C_{16}}{n^\nu}\|u\|_{H_1}, \quad \alpha = n^{-1}, \quad n \in \mathbb{N}. \qquad (7.25)$$

As one can see from (7.25), $\|x_n - x_\xi^*\|_{H_1}$ is of the order $O(n^{-\nu})$, where, unlike the case of (7.20) and (7.21), the exponent $\nu > 0$ does not have an upper bound. Thus, the convergence rate of process (6.1) can be arbitrarily fast provided that initial discrepancy $\xi - x_\xi^*$ has the corresponding level of 'smoothness' ν. In case of approximations (5.5), regardless of the value of $\nu > 0$, the above arguments cannot guarantee the order higher than $O(\alpha)$ for $\|x_\alpha - x_\xi^*\|_{H_1}$. In connection with that, method (5.5) is said to have a *saturation* (with respect to the order of approximation that depends on ν) *property*, while method (6.1) is *saturation free*. By (6.18) and (7.25),

$$\|x_n(\tilde{f}) - x_\xi^*\|_{H_1} \leq C_{17} \left(\frac{1}{n^\nu} \|u\|_{H_1} + \sqrt{n}\delta \right). \tag{7.26}$$

Obviously, if $n = n(\delta)$ is chosen to ensure minimum of the right-hand side in (7.26) with respect to $n \in \mathbb{N}$, then the following inequality holds

$$\sup_{\tilde{f}: \|\tilde{f} - f\|_{H_2} \leq \delta} \|x_{n(\delta)}(\tilde{f}) - x_\xi^*\|_{H_1} = C_{18} \|u\|_{H_1}^{1/(2\nu+1)} \delta^{2\nu/(2\nu+1)}. \tag{7.27}$$

7.4 Approximations in case of a perturbed STC

As it has been emphasized above, the problem of finding the element ξ from the ellipsoid S_d (or Σ_d) is hardly simpler than the problem of solving the original equation. This means that it is highly unlikely one can algorithmically guarantee in practice the exact satisfaction of source-type condition (7.3) (or (7.17)) with values of the norms $\|v\|_{H_2}$, $\|u\|_{H_1}$ being under control. It is more realistic to assume that the selected element ξ complies with condition (7.3) (or (7.17)) approximately.

Let us investigate this situation using scheme (5.5) as an example. Suppose the element ξ, instead of (7.3), satisfies the perturbed source-type condition

$$\xi - x_\xi^* = A^* v + w, \quad \|w\|_{H_1} \leq \Delta. \tag{7.28}$$

From geometrical point of view relation (7.28) implies that the element ξ is in the Δ-neighborhood of the ellipsoid S_d. This neighborhood, unlike the ellipsoid S_d itself, is a set with a nonempty interior. Directly from (7.1), taking into the account (5.4), in place of (7.4)–(7.6) one gets the following estimates

$$\|x_\alpha - x_\xi^*\|_{H_1} \leq \frac{\sqrt{\alpha}}{2} \|v\|_{H_2} + \Delta, \tag{7.29}$$

$$\|x_\alpha(\tilde{f}) - x_\xi^*\|_{H_1} \leq \frac{\sqrt{\alpha}}{2} \|v\|_{H_2} + \frac{\delta}{2\sqrt{\alpha}} + \Delta, \tag{7.30}$$

$$\sup_{\tilde{f}: \|\tilde{f} - f\|_{H_2} \leq \delta} \|x_{\alpha(\delta)}(\tilde{f}) - x_\xi^*\|_{H_1} \leq \sqrt{\|v\|_{H_2}\delta} + \Delta. \tag{7.31}$$

Hence, approximations x_α and $x_\alpha(\tilde{f})$ are stable with respect to the perturbations in source-type condition (7.3). If these perturbations are not large, then they do not affect the approximations considerably.

7.5 Accuracy of the estimates

The results, obtained in this chapter, about the convergence rates of methods (5.5) and (6.1) in the presence of noise are based on estimates (7.4), (7.15), (7.20)–(7.21), and (7.25) related to the case of the exact right-hand side in equation (5.2). All the above estimates follow directly from identity (5.10) for the norm of a function of an operator. One can ask a natural question about how accurate these estimates are. In other words, can one get the same kind of estimates under weaker assumptions on the initial discrepancy $\xi - x_\xi^*$ as compared to source-type conditions (7.3), (7.17)? For example, can estimates (7.20)–(7.21) and (7.25) still be satisfied with the exponents $\nu \in (0, 1]$ and $\nu > 0$ respectively, while source-type condition (7.17) is fulfilled with some smaller exponent $\mu < \nu$. Since formula (5.10) provides the exact value of the norm and the subsequent algebraic transformations are also everywhere exact with respect to the orders of α and n on the classes of the source-type solutions, the above estimates are not possible to improve, in general. It is remarkable that they actually remain exact for every particular problem (5.2) as well.

Let us clarify the corresponding result for iterative process (6.1) and source-type condition (7.17). Assume that for iterative process (6.1) with the initial point $x_0 = \xi$ the following estimate of the convergence rate holds

$$\|x_n - x_\xi^*\|_{H_1} \leq C_{19} n^{-\nu} \quad \nu > 0. \tag{7.32}$$

Then for every $\mu \in (0, \nu)$ the inclusion below is satisfied

$$\xi - x_\xi^* \in R((A^*A)^\mu). \tag{7.33}$$

It is easy to verify that

$$R((A^*A)^\nu) \subset R((A^*A)^\mu) \quad \forall \mu \in [0, \nu). \tag{7.34}$$

Thus, estimate (7.25) cannot be fulfilled even in case of an isolated problem, if $\xi - x_\xi^*$ lies in some bigger (as compared to (7.17)) class $R((A^*A)^{\mu_0})$ with $\mu_0 \in (0, \nu)$, and does not belong to $R((A^*A)^\mu)$ with the values $\mu > \mu_0$ close to ν. To summarize, one may conclude that assumption (7.17), sufficient for estimate (7.25) to hold, is also rather close to being the necessary condition for this estimate. In general, however, one cannot claim that estimate (7.32) implies (7.17). In case of scheme (5.5), the inequality

$$\|x_\alpha - x_\xi^*\|_{H_1} \leq C_{20}\alpha^\nu, \quad \nu \in (0, 1),$$

yields (7.33) for all $\mu \in (0, \nu)$. If in the previous estimate $\nu = 1$, i.e.,

$$\|x_\alpha - x_\xi^*\|_{H_1} \leq C_{20}\alpha, \tag{7.35}$$

then $\xi - x_\xi^* \in R(A^*A)$. In other words, the last inclusion is the necessary and sufficient condition for inequality (7.35).

Finally, let us say a few words about the convergence rates for iterations (6.14) combined with a posteriori stopping rule (6.21). One can show that in this case estimate (7.27) is also satisfied under source-type condition (7.17). In the above a priori stopping rule the number $n = n(\delta)$, minimizing the right-hand side of (7.26) (or its upper bound, obtained by replacing $\|u\|_{H_1}$ with d), depends on ν, see Problem 7.3. The a posteriori rule (6.21) is capable of adjusting the stopping time to the existing level of 'smoothness' of the initial discrepancy $\xi - x_\xi^*$. This is an advantage of the a posteriori stopping rule over the a priori ones.

Problems

7.1. Prove relations (7.11) and (7.12).

7.2. Derive estimate (7.21).

7.3. How should one choose a regularization parameter $\alpha = \alpha(\delta)$ to satisfy estimate (7.24)? Answer the same question regarding the function $n = n(\delta)$ and estimate (7.27). Assuming that $\xi \in \Sigma_d$, find the functions $\alpha(\delta)$ and $n(\delta)$ that guarantee the estimates of the same order as in (7.24) and (7.27).

7.4. Assume that series expansion (5.8) of the initial discrepancy $\xi - x_\xi^*$ in eigenvectors of the operator A^*A has a finite number of terms only. Prove that in this case condition (7.3) holds and condition (7.17) is fulfilled for any $\nu > 0$.

7.5. Investigate approximation errors of schemes (5.5) and (6.1) under the assumption that, instead of (7.17), the perturbed source-type condition is satisfied

$$\xi - x_\xi^* = (A^*A)^\nu u + w, \quad u, w \in H_1, \quad \|w\|_{H_1} \le \Delta \quad (\nu > 0).$$

Give analogs of estimates (7.29)–(7.31).

7.6. Prove inclusion (7.34). Show that if $\dim H_1 < \infty$, then, for $\mu \in (0, \nu)$, this inclusion must be replaced with the identity.

7.7. Prove that if the operator A^*A has infinite number of eigenvalues $\lambda_k > 0, k \in \mathbb{N}$ ($\lim_{k \to \infty} \lambda_k = 0$), then inclusion (7.34) is strict. *Hint*: verify that for every sequence $\{M_k\}$, $\lim_{k \to \infty} M_k = \infty$, there exists a sequence $\{a_k\}$ such that $a_k > 0, k \in \mathbb{N}$, and

$$\sum_{k=1}^\infty a_k < \infty, \quad \sum_{k=1}^\infty a_k M_k = \infty.$$

8

Equations with a convex discrepancy functional by Tikhonov's method

8.1 Some difficulties associated with Tikhonov's method in case of a convex discrepancy functional

The case of operator (5.1) is not the only case when Tikhonov's scheme, presented in Chapter 4, generates effective methods for computing approximate solutions to problem (1.1). The key to success of that scheme for equation (5.2) with an affine operator F is the quadratic structure of the discrepancy functional

$$\Phi(x) = \frac{1}{2}\|F(x)\|_{H_2}^2, \quad x \in H_1.$$

With the functional Φ being quadratic, one can calculate the minimizer of Tikhonov's functional 'explicitly' (see formula (5.5)), using the necessary and sufficient conditions of minimum. In this chapter we analyze a more general case, i.e., the case when the functional Φ is convex, but not necessarily quadratic. As before, we assume that $F \in \mathcal{F}(N_1, N_2)$. For equation (5.2), the corresponding functional

$$\Phi(x) = \frac{1}{2}\|Ax - f\|_{H_2}^2$$

is convex and, therefore, the class of operator equations with a convex discrepancy functional Φ is not empty. Obviously, this class is an extension of the class of equations with affine operators $F \in \mathcal{F}(N_1, 0)$. The direct differentiation of Tikhonov's functional

$$\Phi_\alpha(x) = \Phi(x) + \frac{\alpha}{2}\|x - \xi\|_{H_1}^2$$

gives

$$\Phi'_\alpha(x) = \Phi'(x) + \alpha(x - \xi), \quad x \in H_1.$$

In addition to that, since Φ is convex, the following condition holds

$$(\Phi'(x) - \Phi'(y), x - y)_{H_1} \geq 0 \quad \forall x, y \in H_1.$$

Thus,
$$(\Phi'_\alpha(x) - \Phi'_\alpha(y), x - y)_{H_1} \geq \alpha \|x - y\|^2_{H_1} \quad \forall x, y \in H_1. \tag{8.1}$$

Inequality (8.1) reflects the fact that the functional Φ_α is strongly convex in the space H_1 for any $\alpha > 0$. As it is known from the theory of convex optimization, in this case there exists a unique global minimizer of Φ_α in H_1, x_α, such that
$$x_\alpha \in H_1, \quad \Phi_\alpha(x_\alpha) = \inf_{x \in H_1} \Phi_\alpha(x), \quad \alpha > 0.$$

Using the convexity of the functional Φ, it is not difficult to show that the set $X^*(F)$ of solutions to equation (1.1) is convex and closed. Recall that, by the assumption, $X^*(F) \neq \emptyset$. In our current situation, there is no need to additionally require the weak continuity of the operator F, because the convexity condition on the functional Φ plays the part of this requirement. It follows from the convexity and the closeness of the solution set $X^*(F)$ that, like in case (5.1), there is a unique solution x^*_ξ, nearest to an arbitrarily selected element $\xi \in H_1$:
$$x^*_\xi \in X^*(F), \quad \|x^*_\xi - \xi\|_{H_1} = \mathrm{dist}(\xi, X^*(F)).$$

Hence, relation (5.6) remains valid also for equations with an arbitrary convex functional Φ. Clearly, in this case, it is not possible, in general, to get an 'explicit' formula for x_α, similar to the one in (5.5). However, in the modern optimization theory, numerous effective iterative processes for minimization of strongly convex functionals have been developed [67]. These processes can be used for finding an element x_α, for a fixed $\alpha > 0$, with an arbitrarily high accuracy. While implementing this kind of scheme for the approximation of x^*_ξ, one comes across two typical problems.

The first problem consists in the lack of estimates for the convergence rate of the approximation x_α to x^*_ξ as $\alpha \to 0$. This aspect has already been discussed in the previous chapters.

The second difficulty is encountered due to the fact that iterative processes for minimizing the functional Φ_α get less effective as the regularization parameter $\alpha > 0$ is decreasing, since the above processes for computing the element x_α with a specific accuracy require more and more iterations as $\alpha \to 0$.

8.2 An illustrative example

Let us clarify the above mentioned difficulty using the gradient descent method with a constant step, applied to the functional Φ_α for a fixed regularization parameter $\alpha > 0$, as an example. This method takes the form (compare to (3.1) and (6.1))
$$x_0 \in H_1, \quad x_{n+1} = x_n - \gamma \Phi'_\alpha(x_n), \quad n = 0, 1, \ldots. \tag{8.2}$$

Based on (8.2) and the necessary condition for minimum
$$\Phi'_\alpha(x_\alpha) := \Phi'(x_\alpha) + \alpha(x_a - \xi) = 0,$$

one has

$$\begin{aligned}\|x_{n+1} - x_\alpha\|_{H_1}^2 &= \|x_n - x_\alpha - \gamma(\Phi'_\alpha(x_n) - \Phi'_\alpha(x_\alpha))\|_{H_1}^2 \\ &= \|x_n - x_\alpha\|_{H_1}^2 - 2\gamma(\Phi'_\alpha(x_n) - \Phi'_\alpha(x_\alpha), x_n - x_\alpha)_{H_1} \\ &\quad + \gamma^2 \|\Phi'_\alpha(x_n) - \Phi'_\alpha(x_\alpha)\|_{H_1}^2.\end{aligned} \tag{8.3}$$

Recall that

$$\Phi'(x) = F'^*(x)F(x), \quad x \in H_1.$$

The following estimate holds

$$\begin{aligned}\|\Phi'_\alpha(x_n) - \Phi'_\alpha(x_a)\|_{H_1} &\leq \|\Phi'(x_n) - \Phi'(x_a)\|_{H_1} + \alpha\|x_n - x_\alpha\|_{H_1} \\ &\leq \|F'^*(x_n)\|_{\mathcal{L}(H_2, H_1)} \|F(x_n) - F(x_\alpha)\|_{H_2} \\ &\quad + \|F'^*(x_n) - F'^*(x_\alpha)\|_{\mathcal{L}(H_2, H_1)} \|F(x_\alpha)\|_{H_2} \\ &\quad + \alpha\|x_n - x_\alpha\|_{H_1} \\ &\leq (N_1^2 + N_2\|F(x_\alpha)\|_{H_2} + \alpha) \|x_n - x_\alpha\|_{H_1}.\end{aligned} \tag{8.4}$$

By definition of x_α (see also (4.3)),

$$\Phi_\alpha(x_\alpha) = \frac{1}{2}\|F(x_\alpha)\|_{H_2}^2 + \frac{\alpha}{2}\|x_\alpha - \xi\|_{H_1}^2 \leq \Phi_\alpha(x_\xi^*) = \frac{\alpha}{2}\|x_\xi^* - \xi\|_{H_1}^2.$$

Hence,

$$\|F(x_\alpha)\|_{H_2} \leq \sqrt{\alpha}\|x_\xi^* - \xi\|_{H_1}. \tag{8.5}$$

Inequalities (8.4) and (8.5) imply

$$\begin{aligned}\|\Phi'_\alpha(x_n) - \Phi'_\alpha(x_\alpha)\|_{H_1} &\leq L\|x_n - x_\alpha^*\|_{H_1}, \\ L &= N_1^2 + N_2\|x_\xi^* - \xi\|_{H_1}\sqrt{\alpha} + \alpha.\end{aligned} \tag{8.6}$$

Substituting estimates (8.4) and (8.6) into (8.3) and taking into the account (8.1), one gets

$$\|x_{n+1} - x_\alpha\|_{H_1} \leq \sqrt{1 - 2\gamma\alpha + \gamma^2 L^2}\, \|x_n - x_\alpha\|_{H_1}, \quad n = 0, 1, \ldots. \tag{8.7}$$

One can easily see that for any $\gamma > 0$ the expression under the radical in (8.7) is positive. The step size $\gamma > 0$ must be chosen to make this expression as small as possible. Let us require that

$$1 - 2\gamma\alpha + \gamma^2 L^2 < 1.$$

For this inequality to hold it is sufficient to assume

$$0 < \gamma < \frac{2\alpha}{L^2}. \tag{8.8}$$

If (8.8) is fulfilled, then estimate (8.7) yields

$$\|x_n - x_\alpha\|_{H_1} \leq \|x_0 - x_\alpha\|_{H_1} q^n(\alpha), \quad n = 0, 1, \ldots,$$
$$q(\alpha) = \sqrt{1 - 2\gamma\alpha + \gamma^2 L^2}, \quad q(\alpha) \in (0, 1). \tag{8.9}$$

According to (8.9), the sequence $\{x_n\}$, defined in (8.2), converges to x_α linearly. The quadratic polynomial under the radical in (8.7) attains its minimum with respect to γ when $\gamma = \alpha L^{-2}$. The corresponding minimal value of the polynomial is $1 - \alpha^2 L^{-2}$. Thus, if $\alpha \to 0$, then $q(\alpha)$ in (8.9) converges to 1 for any γ satisfying condition (8.8). Therefore, as the regularization parameter α is decreasing, the number of iterations in (8.2) must go up in order to insure the uniform (with respect to α) accuracy of the approximation of x_α by elements x_n.

Finally, note that if one chooses

$$\gamma = \frac{\varepsilon\alpha}{L^2}, \quad \varepsilon \in (0, 2),$$

then the expression under the radical in (8.7) takes the form

$$1 - 2\gamma\alpha + \gamma^2 L^2 = 1 - C(\varepsilon)\alpha^2, \quad C(\varepsilon) = \frac{\varepsilon(2 - \varepsilon)}{L^2}.$$

Thus, estimate (8.7) can be written as

$$\|x_{n+1} - x_\alpha\|_{H_1} \leq \sqrt{1 - C(\varepsilon)\alpha^2}\,\|x_n - x_\alpha\|_{H_1}.$$

Problems

8.1. Prove that under the assumptions of this chapter the set $X^*(F)$ is convex and closed.

8.2. Using the argument of Chapter 3, investigate gradient scheme (8.2) for a non-convex functional $\Phi(x)$ without the regularity condition. Show that in this case, in addition to the decrease of $\{\Phi_\alpha(x_n)\}$, one can also prove that the iterative sequence $\{x_n\}$ is bounded:

$$\sup_{n \in \mathbb{N}} \|x_n\|_{H_1} \leq D(\alpha) < \infty,$$

provided that the step size γ is chosen appropriately.

8.3. Consider the family of the discrepancy functionals

$$\Phi(x) = \frac{1}{2}\|F(x) - f\|_{H_2}^2, \quad x \in H_1, \tag{8.10}$$

that correspond to the equations $F(x) = f$, $x \in H_1$, for various right-hand sides $f \in H_2$. Prove that if functional (8.10) is convex for all $f \in H_2$, then the operator F is of the form (5.1). *Hint*: Verify that

$$(F'(x) - F'(y))(x - y) = 0 \quad \forall x, y \in H_1$$

and use formula (1.4).

8.4. Let functional (8.10) be convex on H_1 and the element f be given by its δ-approximation $\tilde{f}$ such that $\|\tilde{f} - f\|_{H_2} \leq \delta$. Investigate convergence of the iterations

$$x_{n+1} = x_n - \gamma \tilde{\Phi}'_\alpha(x_n), \quad \tilde{\Phi}_\alpha(x) = \frac{1}{2}\|F(x) - \tilde{f}\|^2_{H_2} + \frac{\alpha}{2}\|x - \xi\|^2_{H_1}.$$

Prove that if $\delta < N_2^{-1}\alpha$, $\alpha > 0$, then the above process converges linearly to the point $\tilde{x}_\alpha$, a global minimizer of the functional $\tilde{\Phi}_\alpha(x)$ on H_1, if $\gamma > 0$ is sufficiently small.

9

Iterative regularization principle

9.1 The idea of iterative regularization

The results, derived in the previous chapters, allow one, in case of a convex discrepancy functional, to get a strongly convergent minimizing sequence directly for that functional. Indeed, take an arbitrary sequence of the regularization parameters $\{\alpha_n\}$ ($\alpha_n > 0$) that converges to zero. Then, by using one of the techniques of convex optimization, find an approximate minimum of a strongly convex functional Φ_{α_n} with an accuracy $\varepsilon_n = \varepsilon(\alpha_n) > 0$. To this end, one has to make some finite number $N(\alpha_n)$ of steps of the method chosen. Denote the point, obtained in the above manner, by $x_{\alpha_n}^{\varepsilon_n}$. Write relations (4.4) and (4.5) for that point:

$$\Phi(x_{\alpha_n}^{\varepsilon_n}) \leq \frac{\alpha_n}{2}\|x^* - \xi\|_{H_1}^2 + \varepsilon_n, \quad \Phi(x) = \frac{1}{2}\|F(x)\|_{H_2}^2, \quad (9.1)$$

$$\|x_{\alpha_n}^{\varepsilon_n} - \xi\|_{H_1}^2 \leq \|x^* - \xi\|_{H_1}^2 + \frac{2\varepsilon_n}{\alpha_n}. \quad (9.2)$$

Following Chapter 4 suppose that

$$\lim_{n\to\infty} \varepsilon_n = \lim_{n\to\infty} \alpha_n = \lim_{n\to\infty} \frac{\varepsilon_n}{\alpha_n} = 0. \quad (9.3)$$

According to (9.2) and (9.3), without loss of generality, one can assume that the sequence $\{x_{\alpha_n}^{\varepsilon_n}\}$ converges weakly in H_1 to some point z^*. However, unlike the case of Chapter 4, in this chapter there is no need to make an assumption about the weak continuity of the operator F, since a continuous convex functional Φ is weakly lower semi-continuous, therefore

$$\Phi(z^*) \leq \liminf_{n\to\infty} \Phi(x_{\alpha_n}^{\varepsilon_n}).$$

From (9.1) and (9.3) it now follows that $\Phi(z^*) = 0$, i.e., $z^* \in X^*(F)$. Repeating the arguments of Chapter 4 and using (9.2), (9.3), one gets

$$\lim_{n\to\infty} \|x_{\alpha_n}^{\varepsilon_n} - x_\xi^*\|_{H_1} = 0.$$

Here x_ξ^* is the element of $X^*(F)$, nearest to ξ. Thus, $\{x_{\alpha_n}^{\varepsilon_n}\}$ is a minimizing sequence for the discrepancy functional that converges strongly to $x_\xi^* \in X^*(F)$. Recall that in our case the set $X^*(F)$ is convex and closed. Besides, by the assumption, $X^*(F) \neq \emptyset$. In the previous chapter we have seen that $N(\alpha_n)$ is unboundedly increasing as $\alpha_n, \varepsilon_n \to 0$. In practice, the above scheme is not convenient due to the fact that it is two-step and also due to the uncertainty with the choice of $N(\alpha_n)$. It turns out, one can generate a sequence $\{x_n\}$, convergent to x_ξ^*, by means of a special one-step iterative process. Let us describe a general scheme for the construction of such processes.

Take some *basic scheme* for the minimization of a strongly convex functional and a sequence $\{\alpha_n\}$ such that

$$0 < \alpha_{n+1} \leq \alpha_n, \quad n = 0, 1, \ldots, \quad \lim_{n \to \infty} \alpha_n = 0. \tag{9.4}$$

Suppose, the point x_n has already been constructed. Then we obtain the next point x_{n+1} by doing a fixed number (independent on n) of steps of the basic iterative method applied to the minimization problem for the Tikhonov functional Φ_{α_n} with the initial point x_n. Once this is done, we repeat the process with $\alpha = \alpha_{n+1}$ and with the initial point x_{n+1}, etc. This algorithm for the construction of convergent minimizing sequences for the discrepancy functional Φ is called the *iterative regularization principle* [10]. Emphasize that for its successful implementation one needs to

a) choose a basic method, acceptable for finding a global minimizer of an arbitrary strongly convex functional, and to fix a number of iterations that one has to perform at every step of the main iterative process; in most cases, one iteration is performed;

b) choose a sequence of the regularization parameters $\{\alpha_n\}$, satisfying conditions (9.4);

c) adjust the inner parameters of the basic method (step size, etc) and the regularization parameter $\alpha_n, n = 0, 1, \ldots$.

9.2 The iteratively regularized gradient method

To analyze the convergence of iteratively regularized algorithms, a general scheme has been developed. We demonstrate how it can be applied by using the gradient descent method with a constant step as the basic method. The resulting iterative process is called the *iteratively regularized gradient method*.

The iteratively regularized gradient method takes the form

$$x_0 \in H_1, \quad x_{n+1} = x_n - \gamma_n \Phi'_{\alpha_n}(x_n)$$
$$= x_n - \gamma_n (F'^*(x_n) F(x_n) + \alpha_n (x_n - \xi)). \tag{9.5}$$

The step size $\gamma_n > 0$, which is an inner parameter of the gradient method, has to be chosen as a function of the regularization parameter α_n. Let $\{\alpha_n\}$ be an arbitrary

9.2 The iteratively regularized gradient method

sequence, satisfying conditions (9.4). As it was shown in Chapter 4,

$$\lim_{n\to\infty} \|x_{\alpha_n} - x_\xi^*\|_{H_1} = 0. \tag{9.6}$$

According to (9.6), convergence of the sequence $\{x_n\}$ to x_ξ^* as $n \to \infty$ would follow from

$$\lim_{n\to\infty} \|x_n - x_{\alpha_n}\|_{H_1} = 0. \tag{9.7}$$

Denote

$$\eta_n = \|x_n - x_{\alpha_n}\|_{H_1}.$$

For the value of η_n a certain difference inequality is fulfilled. In order to drive this inequality, one has to obtain an estimate on the norm of the difference between the elements x_α and x_β, the global minimizers of Tikhonov's functionals Φ_α and Φ_β with two different parameters $\alpha, \beta > 0$. These elements satisfy the operator equations

$$\Phi'_\alpha(x_\alpha) = 0, \quad \Phi'_\beta(x_\beta) = 0,$$

or, in more details,

$$\Phi'(x_\alpha) + \alpha(x_\alpha - \xi) = 0, \tag{9.8}$$
$$\Phi'(x_\beta) + \beta(x_\beta - \xi) = 0. \tag{9.9}$$

Subtracting identity (9.8) from (9.9), one gets

$$(\alpha - \beta)(x_\beta - \xi) = \Phi'(x_\beta) - \Phi'(x_\alpha) + \alpha(x_\beta - x_\alpha).$$

If one scalar multiplies both sides of this equality by $x_\beta - x_\alpha$ and takes into account that convexity of the functional Φ implies

$$(\Phi'(x_\beta) - \Phi'(x_\alpha), x_\beta - x_\alpha)_{H_1} \geq 0,$$

then one concludes

$$(\alpha - \beta)(x_\beta - \xi, x_\beta - x_\alpha)_{H_1} \geq \alpha \|x_\beta - x_\alpha\|_{H_1}^2. \tag{9.10}$$

Applying the Cauchy–Schwartz inequality to the left-hand side of (9.10) and dividing both sides by $\|x_\beta - x_\alpha\|_{H_1}$, one has

$$\|x_\beta - x_\alpha\|_{H_1} \leq \frac{|\alpha - \beta|}{\alpha} \|x_\alpha - \xi\|_{H_1} \quad \forall \alpha, \beta > 0. \tag{9.11}$$

Set $\varepsilon(\alpha) = 0$ in (4.5). Then

$$\|x_\alpha - \xi\|_{H_1} \leq \|x_\xi^* - \xi\|_{H_1} \quad \forall \alpha > 0.$$

9 Iterative regularization principle

Therefore (9.11) yields

$$\|x_\beta - x_\alpha\|_{H_1} \leq \frac{|\alpha - \beta|}{\alpha} \|x_\xi^* - \xi\|_{H_1} \quad \forall \alpha, \beta > 0.$$

Now let us derive a difference inequality for η_n. As a starting point we use the following identity

$$x_{n+1} - x_{\alpha_{n+1}} = (x_n - x_{\alpha_n} - \gamma_n \Phi'_{\alpha_n}(x_n)) + (x_{\alpha_n} - x_{\alpha_{n+1}}). \tag{9.12}$$

Applying (8.7) and (9.11), one obtains from (9.12)

$$\eta_{n+1} = \|x_{n+1} - x_{\alpha_{n+1}}\|_{H_1} \leq \|x_n - x_{\alpha_n} - \gamma_n \Phi'_{\alpha_n}(x_n)\|_{H_1} + \|x_{\alpha_n} - x_{\alpha_{n+1}}\|_{H_1}$$
$$\leq \sqrt{1 - 2\gamma_n \alpha_n + \gamma_n^2 L^2}\, \eta_n + \frac{\alpha_n - \alpha_{n+1}}{\alpha_n} \|x_\xi^* - \xi\|_{H_1}, \tag{9.13}$$

where, according to (8.6) and (9.4),

$$L = N_1^2 + N_2 \|x_\xi^* - \xi\|_{H_1} \sqrt{\alpha_0} + \alpha_0.$$

Inequality (9.13) is satisfied for any choice of a sequence of positive step sizes $\{\gamma_n\}$. If one wants to ensure that η_n converges to zero, then one has to guarantee that the coefficient with η_n in (9.13) is less than 1 at least for sufficiently large values of n. This condition holds if

$$\lim_{n \to \infty} \frac{\gamma_n}{\alpha_n} = 0. \tag{9.14}$$

Since

$$\sqrt{1 - 2\gamma_n \alpha_n + \gamma_n^2 L^2} = \sqrt{1 - \gamma_n \alpha_n \left(2 - \frac{\gamma_n}{\alpha_n} L^2\right)},$$

it follows form (9.14) that for all sufficiently large $n \in \mathbb{N}$, one has

$$\sqrt{1 - 2\gamma_n \alpha_n + \gamma_n^2 L^2} \leq 1 - \frac{1}{2}\gamma_n \alpha_n.$$

Thus, staring with some large number n, for the elements of the sequence $\{\eta_n\}$ the inequality

$$\eta_{n+1} \leq \left(1 - \frac{1}{2}\gamma_n \alpha_n\right) \eta_n + \frac{\alpha_n - \alpha_{n+1}}{\alpha_n} \|x_\xi^* - \xi\|_{H_1} \tag{9.15}$$

is fulfilled. Note that from (9.4) and (9.14) one obtains

$$\lim_{n \to \infty} \gamma_n = 0.$$

In order to complete the argument, we need the following auxiliary result.

9.2 The iteratively regularized gradient method

Lemma ([77]). *Let the sequence of nonnegative numbers $\{\omega_n\}$ be such that*

$$\omega_{n+1} \leq (1-a_n)\omega_n + b_n, \quad n = 0, 1, \ldots, \qquad (9.16)$$

where $\{a_n\}, \{b_n\}$ satisfy the assumptions

$$0 < a_n \leq 1, \quad b_n \geq 0, \quad n = 0, 1, \ldots,$$

$$\sum_{n=0}^{\infty} a_n = \infty, \quad \lim_{n \to \infty} \frac{b_n}{a_n} = 0. \qquad (9.17)$$

Then

$$\lim_{n \to \infty} \omega_n = 0. \qquad (9.18)$$

Indeed, inequality (9.16) yields

$$\omega_{n+1} \leq (1-a_n)(1-a_{n-1})\omega_{n-1} + (1-a_n)b_{n-1} + b_n$$

$$\leq \cdots \leq \prod_{j=0}^{n}(1-a_j)\omega_0 + \sum_{i=0}^{n-1} b_i \prod_{j=i+1}^{n}(1-a_j) + b_n. \qquad (9.19)$$

By (9.17) $\lim_{n\to\infty} \prod_{j=0}^n (1-a_j) = 0$ and $\lim_{n\to\infty} b_n = 0$. Thus we have to show that

$$\lim_{n \to \infty} \sum_{i=0}^{n-1} b_i \prod_{j=i+1}^{n}(1-a_j) = 0. \qquad (9.20)$$

To get (9.20) we will use the representation:

$$\sum_{i=0}^{n-1} b_i \prod_{j=i+1}^{n}(1-a_j) = \sum_{i=0}^{m} b_i \prod_{j=i+1}^{n}(1-a_j) + \sum_{i=m+1}^{n-1} b_i \prod_{j=i+1}^{n}(1-a_j). \qquad (9.21)$$

Here m is an arbitrary integer less than n. The first term in the right-hand side of (9.21) can be estimated as follows

$$\sum_{i=0}^{m} b_i \prod_{j=i+1}^{n}(1-a_j) = \sum_{i=0}^{m} \frac{b_i}{a_i}\left[\prod_{j=i+1}^{n}(1-a_j) - \prod_{j=i}^{n}(1-a_j)\right]$$

$$\leq \zeta\left[\prod_{j=m+1}^{n}(1-a_j) - \prod_{j=0}^{n}(1-a_j)\right] \leq \zeta \exp\left(-\sum_{j=m+1}^{n} a_j\right), \qquad (9.22)$$

where $\zeta = \sup_{n \geq 0} \frac{b_n}{a_n}$. Introduce the notation $\varrho_{m,n} := \max_{m+1 \leq i \leq n-1} \frac{b_i}{a_i}$. For the second term in (9.21) one obtains

$$\sum_{i=m+1}^{n-1} b_i \prod_{j=i+1}^{n}(1-a_j) \leq \varrho_{m,n} \sum_{i=m+1}^{n-1}\left[\prod_{j=i+1}^{n}(1-a_j) - \prod_{j=i}^{n}(1-a_j)\right]$$

$$= \varrho_{m,n}\left[1 - a_n - \prod_{j=m+1}^{n}(1-a_j)\right] \leq \varrho_{m,n}. \qquad (9.23)$$

Therefore

$$\lim_{n\to\infty} \sum_{i=0}^{n-1} b_i \prod_{j=i+1}^{n} (1-a_j) \leq \limsup_{m\to\infty}\left\{\lim_{n\to\infty}\left[\zeta\exp\left(-\sum_{j=m+1}^{n} a_j\right) + \varrho_{m,n}\right]\right\}$$

$$\leq \limsup_{m\to\infty}\left(0 + \sup_{m+1\leq n}\frac{b_n}{a_n}\right) = \limsup_{n\to\infty}\frac{b_n}{a_n} = 0. \quad (9.24)$$

The last conclusion is a consequence of the last assumption in (9.17). From (9.24), (9.20) and (9.19) one concludes that (9.18) holds.

In case of estimate (9.15), requirements (9.17) indicate

$$\sum_{n=0}^{\infty} \alpha_n \gamma_n = \infty, \quad \lim_{n\to\infty} \frac{\alpha_n - \alpha_{n+1}}{\alpha_n^2 \gamma_n} = 0. \quad (9.25)$$

It is possible to satisfy conditions (9.4), (9.14) and (9.25) by choosing sequences $\{\alpha_n\}$, $\{\gamma_n\}$ that slowly converge to zero. One can take, for example,

$$\alpha_n = \frac{\alpha_0}{(n+1)^p}, \quad \gamma_n = \frac{\gamma_0}{(n+1)^{p+q}}, \quad n = 0, 1, \ldots, \quad (9.26)$$

where $\alpha_0, \gamma_0 > 0$, $p \in (0, 1)$ and $q > 0$. Then assumptions (9.4) and (9.14) are, obviously, fulfilled. The first condition in (9.25) holds, if

$$2p + q \leq 1. \quad (9.27)$$

One can easily verify that if inequality (9.27) is strict, then the second condition in (9.25) is also satisfied. Moreover, the following convergence rate can be guaranteed for the sequence $\{\eta_n\}$ in this case:

$$\eta_n = O\left(\frac{1}{n^{1-2p-q}}\right). \quad (9.28)$$

For the sake of simplicity we skip the proof of relation (9.28). One concludes from the above that under assumptions (9.4), (9.14) and (9.25), process (9.5) converges to x_ξ^*, i.e., the identity

$$\lim_{n\to\infty} \|x_n - x_\xi^*\|_{H_1} = 0$$

holds, for any choice of the initial approximation $x_0 \in H_1$.

Problems

9.1. Prove that if $p, q > 0$ and $2p + q < 1$, then sequences (9.26) satisfy the second condition in (9.25).

9.2. Investigate convergence of iteratively regularized gradient process

$$x_0 \in H_1, \quad x_{n+1} = x_n - \gamma_n(F'^*(x_n)(F(x_n) - \tilde{f}) + \alpha_n(x_n - \xi)) \quad (9.29)$$

for solving the equation $F(x) = f$ ($x \in H_1$) when, instead of the exact right-hand side, its δ-approximation is available

$$\|\tilde{f} - f\|_{H_2} \leq \delta$$

(see Problem 8.4). How should one stop process (9.29) in order to get an approximation $x_{n(\delta)}$ that converges to the solution as $\delta \to 0$?

9.3. Let it be known a priori that in problem (1.1) the solution $x_\xi^* \in Q$, where Q is a convex closed set in H_1. Denote by P_Q the projection operator (in general, nonlinear) from H_1 onto the set Q, i.e.,

$$P_Q(x) \in Q, \quad \|x - P_Q(x)\|_{H_1} = \text{dist}(x, Q) \quad \forall x \in H_1.$$

Investigate convergence of iteratively regularized gradient projection scheme

$$x_0 \in H_1, \quad x_{n+1} = P_Q(x_n - \gamma_n \Phi'_{\alpha_n}(x_n)), \quad n = 0, 1, \ldots.$$

Hint: Verify the inequality

$$\|P_Q(x) - P_Q(y)\|_{H_1} \leq \|x - y\|_{H_1} \quad \forall x, y \in H_1$$

and use the identity $P_Q(x_\xi^*) = x_\xi^*$.

9.4. In connection with Problem 9.3, write the projection operator P_Q for the sets

$$Q_1 = B(z, r), \quad Q_2 = \{x \in H_1 : (a, x)_{H_1} = b\},$$
$$Q_3 = \{x \in H_1 : (a, x)_{H_1} \leq b\},$$

where $a, z \in H_1$, $b, r \in \mathbb{R}$. In which cases the operator P_Q turns out to be linear? Which of the sets, listed above, have interior points?

9.5. Let $H_1 = L_2(a, b)$ and Q is the set of all functions $x = x(t)$ from the space $L_2(a, b)$ such that $c_1 \leq x(t) \leq c_2$ for almost all $t \in (a, b)$ ($c_1 \leq c_2$). Prove that the set Q is convex and closed. Write the projection operator onto this set. How should one change the formula for P_Q, if the set Q is defined by one-sided inequality $x(t) \geq c_1$, or $x(t) \leq c_2$ (for almost all $t \in (a, b)$)? Do the sets Q, considered in this problem, have interior points?

10

The iteratively regularized Gauss–Newton method

10.1 Convergence analysis

The iterative regularization principle turns out to be successful even if applied to equations (1.1) with arbitrary smooth operators F. Below we consider an iteratively regularized version of Gauss–Newton scheme (2.6) ([13], ch. 4). Suppose for now that the operator F is in the class $\mathcal{F}(N_1, N_2)$. Neither convexity of the discrepancy functional, nor regularity of F in a neighborhood of the solution is assumed.

Iterations of the method we are interested in can be obtained as follows. Let $\{\alpha_n\}$ be some sequence of regularization parameters, satisfying the conditions

$$0 < \alpha_{n+1} \leq \alpha_n, \quad n = 0, 1, \ldots, \quad \lim_{n \to \infty} \alpha_n = 0,$$

$$\sup_{n=0,1,\ldots} \frac{\alpha_n}{\alpha_{n+1}} = r < \infty. \tag{10.1}$$

Suppose x_n is the current point of the iterative process we are trying to construct. Take the Tikhonov functional

$$\Phi_{\alpha_n}(x) = \frac{1}{2}\|F(x)\|_{H_2}^2 + \frac{\alpha_n}{2}\|x - \xi\|_{H_1}^2, \quad x \in H_1.$$

Rewrite this functional by replacing the discrepancy part

$$\Phi(x) = \frac{1}{2}\|F(x)\|_{H_2}^2$$

with its approximation

$$\Phi(x_n; x) = \frac{1}{2}\|F(x_n) + F'(x_n)(x - x_n)\|_{H_2}^2$$

(see Chapter 2). Then, instead of Φ_{α_n}, one gets a strongly convex quadratic functional

$$\Phi_{\alpha_n}(x_n; x) = \frac{1}{2}\|F(x_n) + F'(x_n)(x - x_n)\|_{H_2}^2 + \frac{\alpha_n}{2}\|x - \xi\|_{H_1}^2, \quad x \in H_1. \tag{10.2}$$

10.1 Convergence analysis

The only global minimum of functional (10.2) can be written explicitly. Denote this point by x_{n+1} and regard it as the next element of the iterative process. By direct calculations one obtains x_{n+1} in the following form:

$$x_{n+1} = \xi - (F'^*(x_n)F'(x_n) + \alpha_n E_1)^{-1} F'^*(x_n)[F(x_n) - F'(x_n)(x_n - \xi)]. \tag{10.3}$$

Algorithm (10.3) is called the *iteratively regularized Gauss–Newton (IRGN) method*.

For iterations (10.3) to be well-defined, one has to provide the initial point $x_0 \in H_1$. Let x^* be some element of the solution set $X^*(F)$ of equation (1.1). As before, assume this set is nonempty. In the case of an arbitrary operator $F \in \mathcal{F}(N_1, N_2)$ the set $X^*(F)$ is closed, but does not have any special properties like convexity. Still, under certain additional assumptions on x^*, ξ and x_0, process (10.3) converges to x^*, and one can estimate the rate of convergence of $\{x_n\}$ to x^*. From (10.3) one concludes

$$\begin{aligned} x_{n+1} - x^* &= \xi - x^* \\ &\quad - (F'^*(x_n)F'(x_n) + \alpha_n E_1)^{-1} F'^*(x_n)[F(x_n) - F'(x_n)(x_n - \xi)]. \end{aligned} \tag{10.4}$$

Since $F(x^*) = 0$ and, by (1.5),

$$F(x^*) = F(x_n) + F'(x_n)(x^* - x_n) + G(x_n, x^* - x_n),$$

one has

$$\begin{aligned} F(x_n) - F'(x_n)(x_n - \xi) &= F'(x_n)(x_n - x^*) - F'(x_n)(x_n - \xi) \\ -G(x_n, x^* - x_n) &= F'(x_n)(\xi - x^*) - G(x_n, x^* - x_n). \end{aligned}$$

Now one can rewrite identity (10.4) in the form

$$\begin{aligned} x_{n+1} - x^* &= (F'^*(x_n)F'(x_n) + \alpha_n E_1)^{-1} F'^*(x_n) G(x_n, x^* - x_n) \\ &\quad + \alpha_n (F'^*(x_n)F'(x_n) + \alpha_n E_1)^{-1} (\xi - x^*). \end{aligned} \tag{10.5}$$

Suppose the following source-type condition, analogous to (7.3), holds

$$\xi - x^* = F'^*(x^*)v, \quad \|v\|_{H_2} \le d. \tag{10.6}$$

Condition (10.6) indicates that the element ξ in (10.3) belongs to the ellipsoid

$$S_d = \{x \in H_1 : x = x^* + F'^*(x^*)v, \ \|v\|_{H_2} \le d\},$$

which is defined similar to (7.10). Substitution of (10.6) into (10.5) yields

$$\begin{aligned} x_{n+1} - x^* &= (F'^*(x_n)F'(x_n) + \alpha_n E_1)^{-1} F'^*(x_n) G(x_n, x^* - x_n) \\ &\quad + \alpha_n (F'^*(x_n)F'(x_n) + \alpha_n E_1)^{-1} (F'^*(x^*) - F'^*(x_n))v \\ &\quad + \alpha_n (F'^*(x_n)F'(x_n) + \alpha_n E_1)^{-1} F'^*(x_n) v \end{aligned} \tag{10.7}$$

10 The iteratively regularized Gauss–Newton method

Based on (1.6) and (5.22), for the first term in the right-hand side of (10.7) one gets

$$\|(F'^*(x_n)F'(x_n) + \alpha_n E_1)^{-1} F'^*(x_n) G(x_n, x^* - x_n)\|_{H_1} \le \frac{N_2}{4\sqrt{\alpha_n}} \|x_n - x^*\|_{H_1}^2.$$

By (1.3), (2.11), (5.4) and (10.6), the second term is estimates as follows

$$\|(F'^*(x_n)F'(x_n) + \alpha_n E_1)^{-1} (F'^*(x^*) - F'^*(x_n))v\|_{H_1} \le \frac{N_2 d}{\alpha_n} \|x_n - x^*\|_{H_1}.$$

Finally, for the third term, (5.22) and (10.6) imply

$$\|(F'^*(x_n)F'(x_n) + \alpha_n E_1)^{-1} F'^*(x_n) v\|_{H_1} \le \frac{d}{2\sqrt{\alpha_n}}.$$

Combining the above estimates, one derives from (10.7)

$$\|x_{n+1} - x^*\|_{H_1} \le \frac{N_2}{4\sqrt{\alpha_n}} \|x_n - x^*\|_{H_1}^2 + N_2 d \|x_n - x^*\|_{H_1} + \frac{d\sqrt{\alpha_n}}{2}. \tag{10.8}$$

Let us show that, for sufficiently small values of d and $\|x_0 - x^*\|_{H_1}$, $\|x_n - x^*\|_{H_1}$ converges to zero as $n \to \infty$ and the following estimate for the convergence rate holds

$$\|x_n - x^*\|_{H_1} = O(\sqrt{\alpha_n}). \tag{10.9}$$

Denote

$$\gamma_n = \frac{\|x_n - x^*\|_{H_1}}{\sqrt{\alpha_n}}.$$

According to (10.1) and (10.8), one obtains

$$\gamma_{n+1} \le \frac{1}{4} N_2 \sqrt{r} \gamma_n^2 + N_2 d \sqrt{r} \gamma_n + \frac{1}{2} d\sqrt{r}, \quad n = 0, 1, \ldots. \tag{10.10}$$

In order to prove estimate (10.9) it is sufficient to verify that the sequence $\{\gamma_n\}$ is bounded from above. We proceed by induction. Let d satisfy the condition

$$0 < d < \frac{1}{N_2 \sqrt{r}}. \tag{10.11}$$

Assume that $\gamma_0 \le D$, that is,

$$\|x_0 - x^*\|_{H_1} \le D\sqrt{\alpha_0}.$$

Below we make our requirements on the choice of the constant D more specific. For now, suppose that D is a positive number and that for some $n \ge 0$, $\gamma_n \le D$. Let us prove that in this case $\gamma_{n+1} \le D$. Indeed, (10.10) implies

$$\gamma_{n+1} \le \frac{1}{4} N_2 \sqrt{r} D^2 + N_2 d \sqrt{r} D + \frac{1}{2} d\sqrt{r}. \tag{10.12}$$

The value of the right-hand side in (10.12) does not exceed D if

$$\frac{1}{4}N_2\sqrt{r}D^2 + \frac{1}{2}d\sqrt{r} \leq (1 - N_2 d\sqrt{r})D. \qquad (10.13)$$

Note, that by (10.11) the right-hand side of inequality (10.13) is positive. For the sake of brevity, denote

$$q = N_2 d \sqrt{r},$$

and replace condition (10.11) with a stronger one:

$$q + \frac{1}{\sqrt{2}}\sqrt{q}\sqrt[4]{r} < 1. \qquad (10.14)$$

Obviously, inequality (10.14) holds if d is rather small. The direct calculations show that under condition (10.14) the values of D, satisfying (10.13), fill out a segment $[D_1, D_2]$, where $0 < D_1 < D_2$. If one takes $D \in [D_1, D_2]$, then $\gamma_{n+1} \leq D$. Following the induction principle, one concludes that $\gamma_n \leq D$ for all $n = 0, 1, \ldots$, i.e.,

$$\|x_n - x^*\|_{H_1} \leq D\sqrt{\alpha_n}, \quad n = 0, 1, \ldots. \qquad (10.15)$$

According to the remark made in the beginning of Chapter 1, our initial requirement, $F \in \mathcal{F}(N_1, N_2)$, can be relaxed essentially. By (10.1) and (10.15), all iterative points x_n are contained in the ball $B(x^*, D\sqrt{\alpha_0})$. Therefore it is sufficient to assume that F satisfies condition (1.20) and the following inequality holds

$$D\sqrt{\alpha_0} \leq R.$$

10.2 Further properties of IRGN iterations

Let us discuss some additional results that one can obtain by studying convergence of iterations (10.3).

1) Assume that the element ξ is chosen in such a way that the distance between ξ and the ellipsoid S_d is greater than zero. Namely, let instead of exact source-type condition (10.6) the approximate condition

$$\xi - x^* = F'^*(x^*)v + w, \quad \|v\|_{H_2} \leq d, \quad \|w\|_{H_1} \leq \Delta, \qquad (10.16)$$

be satisfied. The value of Δ in (10.16) is the error, with which condition (10.6) holds. If we use the same argument as the one applied to derive (10.8) and (10.10), but with (10.6) being replaced by (10.16), then, instead of (10.10), we get the following difference inequality

$$\gamma_{n+1} \leq \frac{1}{4}N_2\sqrt{r}\gamma_n^2 + N_2 d\sqrt{r}\gamma_n + \frac{1}{2}d\sqrt{r} + \frac{\Delta\sqrt{r}}{\sqrt{\alpha_n}}, \quad n = 0, 1, \ldots. \qquad (10.17)$$

Since the last term in the right-hand side of estimate (10.17) increases unboundedly as $n \to \infty$, in general, one cannot guarantee that the values γ_n remain uniformly bounded with respect to n. However, it follows from (10.17), that for the numbers n such that

$$\frac{\Delta}{\sqrt{\alpha_n}} \leq \frac{1}{2}d, \tag{10.18}$$

the relation, similar to (10.12), is fulfilled

$$\gamma_{n+1} \leq \frac{1}{4}N_2\sqrt{r}D^2 + N_2 d\sqrt{r}D + d\sqrt{r}. \tag{10.19}$$

Assume additionally that

$$\frac{\Delta}{\sqrt{\alpha_0}} \leq \frac{1}{2}d, \tag{10.20}$$

and define the number $n(\Delta)$ by the identity

$$n(\Delta) = \max\left\{n = 1, 2, \ldots : \frac{\Delta}{\sqrt{\alpha_{n-1}}} \leq \frac{1}{2}d\right\}. \tag{10.21}$$

(10.1) together with (10.20) imply that the number $n(\Delta)$, satisfying (10.21), is always finite and $n \in \mathbb{N}$. Besides,

$$\lim_{\Delta \to 0} n(\Delta) = \infty. \tag{10.22}$$

According to conditions (10.1) and (10.21), inequality (10.18) holds for the numbers $n = 0, 1, \ldots, n(\Delta) - 1$ precisely. For the same numbers, relation (10.19) is satisfied. Repeating the previous arguments with inequality (10.14) being replaced by a stronger one

$$q + \sqrt{q}\sqrt[4]{r} < 1,$$

one gets estimate (10.15), which is fulfilled for the numbers $n = 0, 1, \ldots, n(\Delta)$ only in this case. The iterative point $x_{n(\Delta)}$ can now be taken as an approximation to the solution x^*. Setting $n = n(\Delta)$ in estimate (10.15), one gets

$$\|x_{n(\Delta)} - x^*\|_{H_1} \leq D\sqrt{\alpha_{n(\Delta)}}. \tag{10.23}$$

Here $D \in [\tilde{D}_1, \tilde{D}_2]$ $(0 < \tilde{D}_1 < \tilde{D}_2)$; $\tilde{D}_1, \tilde{D}_2$ are the roots of the quadratic equation

$$\frac{1}{4}N_2\sqrt{r}D^2 + N_2 d\sqrt{r}D + d\sqrt{r} = D.$$

We point out that, by (10.22), the right-hand side of estimate (10.23) converges to zero as $\Delta \to 0$. This statement can be specified as follows. From (10.21) one has

$$\frac{\Delta}{\sqrt{\alpha_{n(\Delta)}}} > \frac{1}{2}d.$$

10.2 Further properties of IRGN iterations

Therefore

$$\sqrt{\alpha_{n(\Delta)}} < \frac{2\Delta}{d},$$

and (10.23) yields

$$\|x_{n(\Delta)} - x^*\|_{H_1} < \frac{2D\Delta}{d}.$$

As one can see, the element $x_{n(\Delta)}$ depends stably on the error in source-type condition (10.6). The urgency in the replacement of original source-type condition (10.6) with relaxed condition (10.16) is justified by the same argument that motivates transition from (7.3) to (7.28) in the linear case. Namely, for irregular equation (1.1), the interior of the ellipsoid S_d is empty. A typical example is an equation with an operator F that has a compact derivative $F'(x)$ ($x \in H_1$) for $\dim H_1 = \infty$ (see Problem 1.12). Unlike (10.6), condition (10.16) requires to choose the element ξ form a Δ-neighborhood of S_d that has a nonempty interior.

2) The convergence analysis of process (10.3) can also be done under a stronger, as compared to (10.6), source-type condition

$$\xi - x^* = (F'^*(x^*)F'(x^*))^\nu u, \quad u \in H_1. \tag{10.24}$$

Here $\nu \geq 1/2$ and $\|u\|_{H_1} \leq d$. Assumption (10.24) is more restrictive than (10.6), since

$$R((F'^*(x^*)F'(x^*))^{1/2}) = R(F'^*(x^*))$$

and the following inclusion holds (see (7.34))

$$R((F'^*(x^*)F'(x^*))^\nu) \subset R((F'^*(x^*)F'(x^*))^{1/2}) \quad \forall \nu > 1/2,$$

which turns out to be strict in most cases. The key to the analysis of iterations (10.3) under condition (10.24) is an identity, similar to (10.7), but the required estimates are more technical if one uses (10.24). In place of (10.9), provided that certain limitations on the values of d and $\|x_0 - x^*\|_{H_1}$ are imposed, one gets the following estimate (see [13], ch. 4, §1):

$$\|x_n - x^*\|_{H_1} = \begin{cases} O(\alpha_n^\nu), & 1/2 \leq \nu \leq 1, \\ O(\alpha_n), & \nu \geq 1. \end{cases} \tag{10.25}$$

The analogy of estimates (10.25) and (7.20)–(7.21) is obvious. However, in case of (10.25) the exponent in the source-type condition ν is bounded from below by the value $1/2$. Together with (10.24), one can consider perturbed source-type conditions of the form ([13], ch. 4, §1)

$$\xi - x^* = (F'^*(x^*)F'(x^*))^\nu u + w, \quad \|u\|_{H_1} \leq d, \quad \|w\|_{H_1} \leq \Delta. \tag{10.26}$$

3) Our statement that source-type condition (10.24) implies estimate (10.25) admits a reversion in the following sense. One can show that ([13], ch. 4, §2) if process (10.3) has the convergence rate

$$\|x_n - x^*\|_{H_1} = O(\alpha_n^\nu), \quad n = 0, 1, \ldots \quad (1/2 < \nu \le 1),$$

then for all $\mu \in [0, \nu)$ the inclusion below holds

$$\xi - x^* \in R((F'^*(x^*)F'(x^*))^\mu).$$

Hence, condition (10.24), sufficient for the realization of estimate (10.25), is quite close to being the necessary one. In Chapter 7 we have already come across converse theorems of this kind, related to various methods for solving linear irregular equations (5.2).

4) Up until now it was assumed that the operator F, which defines equation (1.1), is known exactly. Let us see how the above arguments are to be changed if the operator F is no longer accurate. To this end, consider the equation

$$F(x) = f, \quad x \in H_1 \quad (f \in H_2) \tag{10.27}$$

with the operator $F \in \mathcal{F}(N_1, N_2)$. For this equation, process (10.3) takes the form

$$x_{n+1} = \xi - (F'^*(x_n)F'(x_n) + \alpha_n E_1)^{-1} F'^*(x_n)[F(x_n) - f - F'(x_n)(x_n - \xi)]. \tag{10.28}$$

Suppose that the exact right-hand side of equation (10.27) is unknown and, instead, its approximation $\tilde{f} \in H_2$ is available such that

$$\|\tilde{f} - f\|_{H_2} \le \delta. \tag{10.29}$$

Naturally, in this case, one calculates an approximate value of x^* using the following iterative process

$$x_{n+1} = \xi - (F'^*(x_n)F'(x_n) + \alpha_n E_1)^{-1} F'^*(x_n)[F(x_n) - \tilde{f} - F'(x_n)(x_n - \xi)], \tag{10.30}$$

which is obtained from (10.28) by replacing f with $\tilde{f}$. Let a solution x^* to equation (10.27) satisfy perturbed source-type condition (10.16) with some element $\xi \in H_1$. By repeating arguments (10.4)–(10.7) and taking into account (10.29), one gets the estimate

$$\|x_{n+1} - x^*\|_{H_1} \le \frac{N_2}{4\sqrt{\alpha_n}} \|x_n - x^*\|_{H_1}^2$$
$$+ N_2 d \|x_n - x^*\|_{H_1} + \frac{d\sqrt{\alpha_n}}{2} + \Delta + \frac{\delta}{2\sqrt{\alpha_n}} \tag{10.31}$$

in place of (10.8). Based on (10.31), one concludes that the following inequality is now fulfilled, instead of (10.17),

$$\gamma_{n+1} \leq \frac{1}{4} N_2 \sqrt{r} \gamma_n^2 + N_2 d \sqrt{r} \gamma_n + \frac{1}{2} d \sqrt{r} + \frac{\Delta \sqrt{r}}{\sqrt{\alpha_n}} + \frac{\delta \sqrt{r}}{2\alpha_n}, \quad n = 0, 1, \ldots. \tag{10.32}$$

Just like in case of (10.20) assume that

$$\frac{\Delta}{\sqrt{\alpha_0}} + \frac{\delta}{2\alpha_0} \leq \frac{1}{2} d.$$

Since $\gamma_n \leq D$, inequality (10.32) implies that for all numbers $n = 0, 1, \ldots, n(\delta, \Delta) - 1$, where

$$n(\delta, \Delta) = \max \left\{ n = 1, 2, \ldots : \frac{\Delta}{\sqrt{\alpha_{n-1}}} + \frac{\delta}{2\alpha_{n-1}} \leq \frac{1}{2} d \right\}, \tag{10.33}$$

relation (10.19) holds. Take $x_{n(\delta,\Delta)}$ as an approximation to the solution x^*. Arguing as above, one derives the estimate

$$\|x_{n(\delta,\Delta)} - x^*\|_{H_1} \leq D \sqrt{\alpha_{n(\delta,\Delta)}}. \tag{10.34}$$

We point out that $n(\delta, \Delta) \in \mathbb{N}$ and

$$\lim_{\delta, \Delta \to 0} n(\delta, \Delta) = \infty.$$

Therefore the right-hand side of estimate (10.34) converges to zero as $\delta, \Delta \to 0$. The number $n(\delta, \Delta)$, defined by formula (10.33), has a meaning of a stopping rule for iterative process (10.30). Estimate (10.34) indicates that method (10.28) is stable with respect to noise in f and with respect to perturbations in source-type condition (10.6), provided the iterations are terminated at a suitable moment.

As before, in place of inclusion $F \in \mathcal{F}(N_1, N_2)$, it is sufficient to require the fulfilment of condition (1.20) in the ball $B(x^*, R)$ with radius $R \geq D \sqrt{\alpha_0}$.

10.3 A unified approach to the construction of iterative methods for irregular equations

In conclusion of this chapter we describe a unified approach to the construction of iterative methods for irregular equations (1.1). As a starting point for the construction of the next approximate element x_{n+1}, given the element x_n, one can take the linearized equation (compare to (1.9))

$$F(x_n) + F'(x_n)(x - x_n) = 0, \quad x \in H_1. \tag{10.35}$$

As it was mentioned in Chapter 1, if the original equation is not regular, then linearized equation (10.35) may not be solvable, or it may have a set of solutions unstable with respect to small variations of the operator F. The same applies to the least square solution of (10.35) (see (2.4)). Nevertheless, *formally*, one can still use a number of methods for *solving* (10.35), among which there are Tikhonov method (5.3) and iterative gradient method (6.1) considered before. Since solvability of (10.35) is not guaranteed, it does not make much of a sense to try to find the exact solution of (10.35) with $\alpha \to 0$ in Tikhonov's method, or $n \to \infty$ in the gradient method. It is worthwhile to stop at some positive value $\alpha = \alpha_n > 0$, or at some iteration of process (6.1) with a finite number $n < \infty$. Let us look closely at Tikhonov method (5.3) with $\alpha = \alpha_n$ applied to equation (10.35). Rewrite this equation in the form

$$F'(x_n)x = F'(x_n)x_n - F(x_n), \quad x \in H_1. \tag{10.36}$$

Assuming in (5.3), or (5.4), $A = F'(x_n)$, $f = F'(x_n)x_n - F(x_n)$, by direct calculations one gets

$$x_{\alpha_n} = \xi - (F'^*(x_n)F'(x_n) + \alpha_n E_1)^{-1} F'^*(x_n)[F(x_n) - F'(x_n)(x_n - \xi)]. \tag{10.37}$$

As one can see, the approximation x_{α_n} is equal to the element x_{n+1} in iterative process (10.3). This alternative argument for obtaining iterations (10.3) allows one to derive other methods in a similar fashion. For example, one can apply a finite number of steps of gradient method (6.1) to equation (10.36) and take the resulting point as the next approximation x_{n+1} in the basic iterative process. It is reasonable to assume that the number of gradient steps for solving linearized equation (10.36) should increase as n goes up. One can make exactly n steps of (6.1) at the nth exterior iteration ($n \geq 1$). In this case, according to (6.7), one has

$$\begin{aligned} x_{n+1} &= [E_1 - \Theta(F'^*(x_n)F'(x_n), \alpha_n) F'^*(x_n) F'(x_n)]\xi \\ &\quad + \Theta(F'^*(x_n)F'(x_n), \alpha_n) F'^*(x_n)(F'(x_n)x_n - F(x_n)) \\ &= \xi - \Theta(F'^*(x_n)F'(x_n), \alpha_n) F'^*(x_n)[F(x_n) - F'(x_n)(x_n - \xi)]. \end{aligned} \tag{10.38}$$

Here $\alpha_n = n^{-1}$ and the function $\Theta(\lambda, \alpha)$ is defined in (6.6).

Is is not difficult to verify, that iterative method (10.3), which was derived before, can also be written in form (10.38), though the generating function $\Theta(\lambda, \alpha)$ must be given by (5.11). Thus, (10.38) turns out to be a rather general scheme that includes various iterative methods for solving irregular equations (1.1). The results listed above in 1)–4) have their natural analogies for generating function (6.6), as well as for many other generating functions Θ (see [13], ch. 4, §1).

10.4 The reverse connection control

In this section, the possibility of using the *reverse connection* control for iteratively regularized Gauss–Newton-type algorithms is analyzed. It is shown that for the *generalized Gauss–Newton iterations* the diameter of the set containing the control elements can tend to ∞ as $n \to \infty$.

10.4 The generalized Gauss–Newton method

Consider iterative method (10.38) for solving the equation

$$F(x) = 0, \quad F : H_1 \to H_2, \tag{10.39}$$

in the presence of noise

$$x_{n+1} = \xi_n - \Theta(F_\delta'^*(x_n) F_\delta'(x_n), \alpha_n) F_\delta'^*(x_n) \{F_\delta(x_n) - F_\delta'(x_n)(x_n - \xi_n)\}. \tag{10.40}$$

We refer the reader to [13] and [16] for a detailed convergence analysis of algorithm (10.40). In this algorithm, $\{\alpha_n\}$ is a regularization sequence that converges to 0 as $n \to \infty$ and $\Theta(\lambda, \alpha)$ is a generating function. It has been shown in [13] that some choices of $\Theta(\lambda, \alpha)$ may result in better convergence rates than those obtained by IRGN method (10.3), which corresponds to $\Theta(\lambda, \alpha) = \frac{1}{\lambda + \alpha}$.

Starting with [11], the following source condition has been imposed to carry out the convergence analysis of (10.40):

$$x^* - \xi \in \varphi(F'^*(x^*)F'(x^*))S, \quad S := \{v \in H_1, \|v\|_{H_1} \le \varepsilon\}. \tag{10.41}$$

Normally, one takes [13]

$$\varphi(t) = t^\mu, \quad \mu \ge \frac{1}{2}, \tag{10.42}$$

though

$$\varphi(t) = t^\mu, \quad 0 \le \mu < \frac{1}{2}, \tag{10.43}$$

and

$$\varphi(t) = \begin{cases} \left(-\ln \frac{t}{e\, t_0}\right)^{-p}, & 0 < t \le t_0, \\ 0, & t = 0 \end{cases} \tag{10.44}$$

($t_0 \ge \|F'(\cdot)\|^2$, $p > 0$, and e is Euler's constant) can also be used if combined with certain nonlinearity conditions [46].

The convergence theorems based on assumption (10.41) require ε to be rather small. However, the practical implementation of iterative process (10.40) reveals a much better tolerance of (10.40) to the choice of ξ than one could expect from the results of the above research. Hence it seems reasonable to further explore the stabilizing properties of algorithm (10.41), and in this section we consider the generalized Gauss–Newton method in the following form

$$x_{n+1} = \xi_n - \Theta(F_\delta'^*(x_n) F_\delta'(x_n), \alpha_n) F_\delta'^*(x_n) \{F_\delta(x_n) - F_\delta'(x_n)(x_n - \xi_n)\}. \tag{10.45}$$

It will be shown here that if one chooses ξ_n by means of undetermined reverse connection, i.e., from the set depending on the current iteration point x_n

$$x^* - \xi_n \in \varphi(F'^*(x^*)F'(x^*))S_n, \tag{10.46}$$

then the radius of S_n (as opposed to ε in (10.41)) can tend to ∞ as $n \to \infty$. Hence the diameter of the set containing elements ξ_n, which guarantee convergence of algorithm (10.40), can become arbitrarily large as iterations progress. Suppose that inequality

$$\|F_\delta(x^*)\|_{H_2} \leq \delta_1 \qquad (10.47)$$

holds and for any $x, y \in \mathcal{B}_\zeta(x^*)$

$$\|F'_\delta(x)\|_{H_2} \leq N, \quad \|F'_\delta(x) - F'_\delta(y)\|_{H_2} \leq L\|x - y\|_{H_1}, \quad N, L \geq 0, \quad (10.48)$$

with ζ defined in (10.65) below. Let the regularizing sequence $\{\alpha_n\}$ and the generating function $\Theta(\lambda, \alpha)$ satisfy the following assumptions

$$1 \leq \frac{\alpha_n^p}{\alpha_{n+1}^p} \leq R, \quad n = 0, 1, 2, \ldots, \qquad (10.49)$$

$$\sup_{\lambda \in [0, N^2]} |\Theta(\lambda, \alpha)\sqrt{\lambda}| \leq \frac{C_1}{\sqrt{\alpha}}, \qquad (10.50)$$

$$\sup_{\lambda \in [0, N^2]} |1 - \Theta(\lambda, \alpha)\lambda|\lambda^p \leq C_2(p)\alpha^p, \quad p \in [0, p_0), \quad p_0 > \frac{1}{2}. \qquad (10.51)$$

By default, it is assumed that $\Theta(\lambda, \alpha)$ is defined for all values of λ and α required to construct iterations (10.45).

For $\Theta(\lambda, \alpha) = \frac{1}{\lambda + \alpha}$, inequality (10.50) is fulfilled with $C_1 = \frac{1}{2}$, and inequality (10.51) holds for $p \in [0, 1]$ with $C_2(p) = p^p(1 - p)^{1-p}$. For some other functions $\Theta(\lambda, \alpha)$, the admissible interval $[0, p_0)$ may be infinite. The specifics of the constants C_1 and C_2 in (10.50) and (10.51) will no longer be relevant to our future reasoning.

To study convergence of (10.45), one has to select a function φ in source condition (10.46). We'll use a modified version of (10.42) in the form

$$x^* - \xi_n \in (F'^*(x_n)F'(x_n))^\mu w_n + v_n, \qquad (10.52)$$

with

$$\frac{1}{2} \leq \mu \leq p_0, \quad \|w_n\|_{H_1} \leq r_n \leq \frac{M}{\alpha_n^k}, \quad 0 \leq k \leq \mu, \quad \text{and} \quad \|v_n\|_{H_1} \leq \delta_2, \qquad (10.53)$$

where M is some positive constant. Assume that, as in the previous sections, x^* is a solution to the exact equation $F(x) = 0$, which is to be approximated by the iterative process. From (10.47) and condition (10.48), one obtains

$$F'^*_\delta(x_n)F_\delta(x_n) = F'^*_\delta(x_n)\{F_\delta(x^*) + F'_\delta(x_n)(x_n - x^*) + \mathcal{G}_n\},$$

$$\|\mathcal{G}_n\|_{H_2} \leq \frac{L}{2}\|x_n - x^*\|^2_{H_1}. \qquad (10.54)$$

10.4 The generalized Gauss–Newton method

By (10.45) and (10.54), it follows that

$$
\begin{aligned}
x_{n+1} - x^* = & -\Theta(F_g'^*(x_n)F_g'(x_n), \alpha_n) F_g'^*(x_n)\{F_8(x^*) + \mathcal{G}_n\} \\
& - [I - \Theta(F_g'^*(x_n)F_g'(x_n), \alpha_n) F_g'^*(x_n)F_g'(x_n)] \\
& \times \{(F_g'^*(x_n)F_g'(x_n))^\mu w_n + v_n\}
\end{aligned}
\tag{10.55}
$$

Taking into account (10.53)–(10.55), one concludes

$$
\|x_{n+1} - x^*\|_{H_1} \leq \frac{C_1 L}{2\sqrt{\alpha_n}} \|x_n - x^*\|_{H_1}^2 + \frac{C_1 \delta_1}{\sqrt{\alpha_n}} + C_2 M \alpha_n^{\mu-k} + C_2 \delta_2. \tag{10.56}
$$

In order to study the behavior of solutions to (10.56) as $n \to \infty$, we introduce a new variable

$$
\gamma_n := \frac{\|x_n - x^*\|_{H_1}}{\alpha_n^p}, \quad \text{where} \quad \frac{1}{2} \leq p \leq \mu - k. \tag{10.57}
$$

Estimates (10.49) and (10.56) yield

$$
\gamma_{n+1} \leq \frac{C_1 L R}{2} \alpha_n^{p-\frac{1}{2}} \gamma_n^2 + \frac{C_1 \delta_1 R}{\alpha_n^{p+\frac{1}{2}}} + C_2 M R \alpha_n^{\mu-k-p} + \frac{C_2 \delta_2 R}{\alpha_n^p}. \tag{10.58}
$$

Assume that n_g^* is the minimal n with

$$
\max\left\{\frac{C_1 \delta_1}{\alpha_n^{p+\frac{1}{2}}}, \frac{C_2 \delta_2}{\alpha_n^p}\right\} > \frac{1}{2} C_2 M \alpha_0^{\mu-k-p}. \tag{10.59}
$$

Let us verify that solutions to (10.58) are bounded for any $n < n_g^*$. Indeed, inequalities (10.57) and (10.59) imply that the coefficients in the right-hand side of (10.58) are bounded. By (10.49), the sequence $\{\alpha_n\}$ approaches zero monotonically. Thus sufficient conditions of boundedness of solutions to the inequality

$$
\gamma_{n+1} \leq \frac{C_1 L R}{2} \alpha_0^{p-\frac{1}{2}} \gamma_n^2 + 2C_2 M R \alpha_0^{\mu-k-p} \tag{10.60}
$$

will also guarantee boundedness of solutions to (10.58).

To show that solutions to (10.60) are bounded, we proceed by induction. Suppose for some $l > 0$ and $n_g^* > n \geq 0$, one has

$$
\gamma_k \leq l, \quad k = 0, \ldots, n. \tag{10.61}
$$

The element γ_{n+1} will satisfy the same inequality if the following condition holds

$$
g_2(l) := \frac{C_1 L R}{2} \alpha_0^{p-\frac{1}{2}} l^2 - l + 2C_2 M R \alpha_0^{\mu-k-p} \leq 0. \tag{10.62}
$$

This inequality will be fulfilled if $l > 0$ is the larger root of the quadratic polynomial $g_2(l)$. Under the assumption

$$
4 C_1 C_2 L M R^2 \alpha_0^{\mu-k-\frac{1}{2}} < 1, \tag{10.63}
$$

this root necessarily exists and equals

$$l^* := \frac{1 + \sqrt{1 - 4C_1C_2LMR^2\alpha_0^{\mu-k-\frac{1}{2}}}}{C_1LR\alpha_0^{p-\frac{1}{2}}}. \tag{10.64}$$

Condition (10.63) can be satisfied for any L and M if α_0 is chosen appropriately. In particular, for $\alpha_0 < 1$, inequality (10.63) holds for any L and M if μ is large enough.

Estimate (10.63) implies that for the solutions to (10.60) to be bounded, it is sufficient that

$$\|x^* - x_0\| \leq \sqrt{\alpha_0} \, \frac{1 + \sqrt{1 - 4C_1C_2LMR^2\alpha_0^{\mu-k-\frac{1}{2}}}}{C_1LR} := \zeta. \tag{10.65}$$

Thus, undetermined reverse connection (10.52) allows one to significantly relax the source condition in the sense specified above.

To summarize, let conditions (10.48)–(10.53), (10.57), (10.59), (10.63) and (10.65) hold. Then for $\{x_n\}$ defined by (10.45), one has

1) $\quad \dfrac{\|x_n - x^*\|_{H_1}}{\theta_n^p} \leq l^* \quad \forall n \leq n_\delta^*, \tag{10.66}$

and

2) $\quad \|x_{n_\delta^*} - x^*\|_{H_1} = O(\delta_1^{1/2} + \delta_2). \tag{10.67}$

Problems

10.1. Prove that identity (10.3) defines minimum of functional (10.2).

10.2. Derive identity (10.5).

10.3. Prove inequality (10.17).

10.4. Derive estimate (10.25) under source-type condition (10.24) with the exponent $\nu = 1$.

10.5. Prove identity (10.37).

10.6. Derive estimates, similar to (10.15) and (10.23), for iterative process (10.38) with generating function (6.6).

10.7. Give explicit formulas for $D_1, D_2, \tilde{D}_1, \tilde{D}_2$. Can the constant D in (10.15) and (10.23) be chosen from the conditions $D \in (0, D_1)$ and $D \in (0, \tilde{D}_1)$ respectively?

10.8. Verify that method (10.38), where the generating function is given by (5.11), coincides with (10.3).

10.4 The generalized Gauss–Newton method

10.9. Suppose that a solution x^* to equation (1.1) is contained in a convex, closed set $Q \subset H_1$. Investigate convergence of the iterative process

$$x_{n+1} = P_Q\{\xi - (F'^*(x_n)F'(x_n) + \alpha_n E_1)^{-1} F'^*(x_n)[F(x_n) - F'(x_n)(x_n - \xi)]\}.$$

Analyze its modification for equation (10.27) with an approximate right-hand side.

10.10. Assuming that the operator $F \in \mathcal{F}(N_1, N_2)$ satisfies condition of the uniform regularity (1.15), consider Newton's method for equation (10.27) with a perturbed right-hand side

$$x_{n+1} = x_n - F'(x_n)^{-1}(F(x_n) - \tilde{f}), \quad n = 0, 1, \ldots.$$

Show that for $\varepsilon_n = \|x_n - x^*\|_{H_1}$ the inequality below

$$\varepsilon_{n+1} \leq \frac{1}{2} m N_2 \varepsilon_n^2 + m\delta$$

holds. Proceeding by mathematical induction and using the technique for the analysis of difference inequalities developed in this chapter, prove that if

$$C \geq 2m, \quad l \leq \frac{1}{mN_2}, \quad \delta \leq \frac{1}{2mN_2C}, \quad \varepsilon_0 \leq lq + C\delta, \quad q \in (0, 1),$$

then the following estimate

$$\varepsilon_n \leq lq^{2^n} + C\delta, \quad n = 1, 2, \ldots,$$

is fulfilled.

11

The stable gradient method for irregular nonlinear equations

11.1 Solving an auxiliary finite dimensional problem by the gradient descent method

In this chapter we consider one more approach to the construction of stable iterative methods for irregular equations (1.1). Suppose that the set of solutions $X^*(F)$ to equation (1.1) is nonempty, $F \in \mathcal{F}(N_1, N_2)$. Let us start with the equivalent variational problem

$$\min_{x \in H_1} \frac{1}{2} \|F(x)\|^2_{H_2}. \tag{11.1}$$

As we have seen in Chapter 3, the direct application of minimization methods (the gradient method, for example) to problem (11.1) in case of irregular operators F does not result in the construction of convergent iterative processes, in general. Ultimately, this is happening because the space H_1 is infinite dimensional. Indeed, the minimization methods, including the gradient one, generate the sequences, along which the functional is decreasing. However, even when those sequences are minimizing and bounded, there is no guarantee they will converge to the solution set $X^*(F)$, if H_1 is infinite dimensional (see Problem 3.5). The situation of a finite dimensional space H_1 is different. It is easy to verify that, if $\dim H_1 < \infty$ and the minimizing sequence is bounded, then every limiting point is contained in $X^*(F)$, where minimum of functional (11.1) is attained.

A similar conclusion can be made based on the analysis of conditions (3.4) and (3.8) that ensure local convergence of gradient iterations (3.1). It has been already mentioned that these conditions are virtually equivalent to the regularity of equation (1.1), and they cannot be fulfilled in case of compact operators $F'(x)$ in an infinite dimensional space H_1, for example. For such operators $0 \in \sigma(F'^*(x)F'(x))$. Obviously, this inclusion takes place if $\lambda = 0$ is an eigenvalue of the operator $F'^*(x)F'(x)$. But even if $\lambda = 0$ is not an eigenvalue, by compactness of $F'(x)$, there is a sequence of eigenvalues that converges to zero. In case of $\dim H_1 < \infty$, the second option is no longer a possibility, and the only chance for the gradient iterations not to converge locally is when some eigenvalue of $F'^*(x)F'(x)$ is zero. To

11.1 An Auxiliary Finite Dimensional Problem

exclude that, one has to require $0 \notin \sigma(F'^{*}(x)F'(x))$, which can be written in the form

$$N(F'(x)) = \{0\}. \tag{11.2}$$

Relation (11.2) is equivalent to injectivity of the operator $F'(x)$ for $x \in H_1$ under consideration.

Having that in mind, we approximate original problem (11.1) by a similar problem of minimizing the discrepancy of equation (1.1) on a finite dimensional subspace of the initial space H_1. In order to make this approximation more practical, we choose this subspace to be *affine*, i.e.,

$$M_\xi = \{x \in H_1 : x = \xi + y, \ y \in M\}.$$

Here $M \subset H_1$ is a *finite dimensional* linear subspace of H_1, the element $\xi \in H_1$ plays the part of a control parameter. Instead of (11.1), one now has the following finite dimensional problem

$$\min_{x \in M_\xi} \frac{1}{2} \|F(x)\|_{H_2}^2.$$

Denote by P_M the orthogonal projection operator from the space H_1 onto the subspace M. It is convenient to rewrite the last problem as

$$\min_{y \in M} \varphi(y), \quad \varphi(y) = \frac{1}{2} \|F(y + \xi - P_M \xi)\|_{H_2}^2, \quad y \in M. \tag{11.3}$$

Let us view the subspace M as a finite dimensional Euclidean space with the scalar product and the norm of the ambient space H_1. Various methods of finite dimensional optimization can be used for solving problem (11.3), [67]. Consider gradient descent method (3.1) with a constant step size as an example. When applied to (11.3), it takes the form

$$y_0 \in M, \quad y_{n+1} = y_n - \gamma \varphi'(y_n), \quad n = 0, 1, \ldots \quad (\gamma > 0). \tag{11.4}$$

The direct differentiation of the functional $\varphi : M \to \mathbb{R}$ gives

$$\varphi'(y) = P_M F'^{*}(y + \xi - P_M \xi) F(y + \xi - P_M \xi), \quad y \in M. \tag{11.5}$$

Using (11.5), present iterations (11.4) as follows

$$y_{n+1} = y_n - \gamma P_M F'^{*}(y_n + \xi - P_M \xi) F(y_n + \xi - P_M \xi),$$
$$n = 0, 1, \ldots, \quad y_0 \in M. \tag{11.6}$$

Set

$$x_n = y_n + \xi - P_M \xi. \tag{11.7}$$

According to (11.7), one writes iterations (11.6) as

$$x_0 \in M_\xi, \quad x_{n+1} = x_n - \gamma P_M F'^*(x_n) F(x_n), \quad n = 0, 1, \ldots, \tag{11.8}$$

(compare to (3.1)). One can see from (11.8) that all iterative points x_n, $n = 1, 2, \ldots$, are contained in the affine subspace M_ξ. Let $x^* \in X^*(F)$ be some solution to equation (1.1). Apparently, the approximations generated by (11.8), being in the set M_ξ, cannot converge to the solution x^* directly, which, generally speaking, does not belong to M_ξ. Below we prove that under certain assumptions the iterative points x_n approximate the projection $\bar{x}^*$ of the solution x^* onto the subspace M_ξ, and estimate the quality of this approximation. It is clear that $\bar{x}^*$ admits the representation

$$\bar{x}^* = P_M x^* + \xi - P_M \xi. \tag{11.9}$$

Let us show that for

$$\varepsilon_n = \|x_n - \bar{x}^*\|_{H_1},$$

the estimate, whose form is similar to inequality (3.7), holds. Recall, it is assumed that the operator F is from the class $\mathcal{F}(N_1, N_2)$. By (11.8) and the identities

$$F(x_n) = F(\bar{x}^*) + F'(\bar{x}^*)(x_n - \bar{x}^*) + G(\bar{x}^*, x_n - \bar{x}^*),$$
$$x_n - \bar{x}^* = P_M(x_n - \bar{x}^*),$$

one gets

$$\begin{aligned} x_{n+1} - \bar{x}^* &= x_n - \bar{x}^* - \gamma P_M F'^*(\bar{x}^*) F'(\bar{x}^*) P_M(x_n - \bar{x}^*) \\ &+ \gamma P_M [F'^*(\bar{x}^*) - F'^*(x_n)] F'(\bar{x}^*)(x_n - \bar{x}^*) \\ &- \gamma P_M F'^*(x_n) F(\bar{x}^*) - \gamma P_M F'^*(x_n) G(\bar{x}^*, x_n - \bar{x}^*). \end{aligned} \tag{11.10}$$

Equality (11.10) together with (1.2), (1.3), (1.6), (2.11) imply

$$\|x_{n+1} - \bar{x}^*\|_{H_1} \leq \|E_1 - \gamma P_M F'^*(\bar{x}^*) F'(\bar{x}^*) P_M\|_{\mathcal{L}(M,M)} \|x_n - \bar{x}^*\|_{H_1}$$
$$+ \frac{3}{2} \gamma N_1 N_2 \|x_n - \bar{x}^*\|_{H_1}^2 + \gamma N_1 \|F(\bar{x}^*)\|_{H_2}. \tag{11.11}$$

Here it is taken into consideration that $\|P_M\|_{\mathcal{L}(H_1, H_1)} = 1$. According to (1.7), the last term in the right-hand side of (11.11) can be estimated as follows

$$\|F(\bar{x}^*)\|_{H_2} = \|F(\bar{x}^*) - F(x^*)\|_{H_2} \leq N_1 \|\bar{x}^* - x^*\|_{H_1}. \tag{11.12}$$

We point out that $\|\bar{x}^* - x^*\|_{H_1}$ is the distance from the solution we are looking for and the subspace M_ξ. Suppose an upper bound for this distance is available, i.e.,

$$\|\bar{x}^* - x^*\|_{H_1} \leq \Delta. \tag{11.13}$$

The last relation can also be written in the form

$$\|(E_1 - P_M)(\xi - x^*)\|_{H_1} \leq \Delta. \tag{11.14}$$

11.1 An Auxiliary Finite Dimensional Problem

Inequality (11.14) admits the following form

$$\xi - x^* = u + w, \quad u \in M, \quad \|w\|_{H_1} \leq \Delta,$$

which is the analogue of perturbed source-type conditions (10.16) and (10.26), with the inclusion

$$\xi - x^* \in M$$

being the analogue of exact conditions (10.6) and (10.24). The above inclusion shows that the affine subspace M_ξ passes directly through the solution x^*.

Obviously, the problem of finding the element ξ, satisfying the inclusion $\xi - x^* \in M$, is not much simpler than the original equation. Condition (11.14), on the contrary, means that ξ is contained in an infinite set of a cylindrical structure with a nonempty interior. One can expect that, in practice, the problem of 'getting' inside this set, namely, in a Δ-neighborhood of the affine subspace M_{x^*}, can be solved with the help of some a priori information about x^* in such a way that Δ turns out to be rather small.

Combining inequalities (11.12) and (11.13), one obtains

$$\|F(\bar{x}^*)\|_{H_2} \leq N_1 \Delta. \tag{11.15}$$

Let us investigate the first term in the right-hand side of (11.11). The expressions $P_M F'^*(\bar{x}^*) F'(\bar{x}^*) P_M$ and $E_1 - \gamma P_M F'^*(\bar{x}^*) F'(\bar{x}^*) P_M$ are considered here as operators acting from the subspace M to M. Since $\dim M < \infty$, its spectrum $\sigma(P_M F'^*(\bar{x}^*) F'(\bar{x}^*) P_M)$ consists of a finite number of nonnegative eigenvalues. By analogy with (11.2), introduce the following a priori assumption

$$N(F'(\bar{x}^*)) \cap M = \{0\}. \tag{11.16}$$

One can easily see that in this case

$$\min\{\lambda : \lambda \in \sigma(P_M F'^*(\bar{x}^*) F'(\bar{x}^*) P_M)\} = \rho > 0. \tag{11.17}$$

Since

$$\|P_M F'^*(\bar{x}^*) F'(\bar{x}^*) P_M\|_{\mathcal{L}(M,M)} \leq N_1^2, \quad \sigma(P_M F'^*(\bar{x}^*) F'(\bar{x}^*) P_M) \subset [\rho, N_1^2],$$

according to formula (5.10), one has

$$\|E_1 - \gamma P_M F'^*(\bar{x}^*) F'(\bar{x}^*) P_M\|_{\mathcal{L}(M,M)} \leq \max_{\lambda \in [\rho, N_1^2]} |1 - \gamma\lambda|. \tag{11.18}$$

From (11.17) it follows that for any $0 < \gamma < 2/N_1^2$ the inequality

$$\max_{\lambda \in [\rho, N_1^2]} |1 - \gamma\lambda| = q < 1. \tag{11.19}$$

holds. Estimates (11.11), (11.15), (11.18), (11.19) yield

$$\varepsilon_{n+1} \leq q\varepsilon_n + \frac{3}{2}\gamma N_1 N_2 \varepsilon_n^2 + \gamma N_1^2 \Delta, \quad n = 0, 1, \ldots. \tag{11.20}$$

11.2 Investigation of a difference inequality

Let us illustrate the technique for the analysis of difference inequalities of type (11.20). Fix an arbitrary $\tilde{q} \in (q, 1)$ and show that, under certain assumptions, there exist positive constants l and C such that

$$\varepsilon_n \leq l\tilde{q}^n + C\Delta, \quad n = 0, 1, \ldots. \tag{11.21}$$

Proceeding by induction, assume that inequality (11.21) is satisfied for $n = 0$, i.e.,

$$\|x_0 - \bar{x}^*\|_{H_1} \leq l + C\Delta. \tag{11.22}$$

The necessary conditions on constants l and C are specified below. Let inequality (11.21) hold for some number $n \geq 0$. Substituting estimate (11.21) into (11.20) and using (6.30), one gets

$$\varepsilon_{n+1} \leq lq\tilde{q}^n + 3\gamma N_1 N_2(l^2\tilde{q}^{2n} + C^2\Delta^2) + \gamma N_1^2 \Delta + qC\Delta.$$

In order to execute the induction step successfully, it is sufficient to ensure that the right-hand side of the above inequality does not exceed $l\tilde{q}^{n+1} + C\Delta$. Evidently, this property is fulfilled if the following assumptions are satisfied

$$lq\tilde{q}^n + 3\gamma N_1 N_2 l^2 \tilde{q}^{2n} \leq l\tilde{q}^{n+1}, \quad 3\gamma N_1 N_2 C^2 \Delta^2 + \gamma N_1^2 \Delta \leq (1-q)C\Delta. \tag{11.23}$$

By direct calculations one can verify that the first inequality in (11.23) holds when the constant l is chosen from the condition

$$0 < l \leq \frac{\tilde{q} - q}{3 N_1 N_2 \gamma}. \tag{11.24}$$

In turn, the second condition in (11.23) is automatically fulfilled if

$$3\gamma N_1 N_2 C^2 \Delta \leq \frac{1-q}{2}C, \quad \gamma N_1^2 \leq \frac{1-q}{2}C. \tag{11.25}$$

Inequality (11.25) takes place if C satisfies the condition

$$C \geq \frac{2\gamma N_1^2}{1-q}, \tag{11.26}$$

and, besides,

$$\Delta \leq \frac{1-q}{6\gamma N_1 N_2 C}. \tag{11.27}$$

Summarizing the above, one concludes that under assumptions (11.22) and (11.27) with constants l, C, chosen according to (11.24) and (11.26), one has

$$\varepsilon_{n+1} \leq l\tilde{q}^{n+1} + C\Delta,$$

and, as the result, estimate (11.21) holds. From (11.21) one obtains

$$\limsup_{n\to\infty} \|x_n - \bar{x}^*\|_{H_1} = \limsup_{n\to\infty} \varepsilon_n \leq C\Delta.$$

Together with (11.13) this implies

$$\limsup_{n\to\infty} \|x_n - x^*\|_{H_1} \leq (C+1)\Delta. \tag{11.28}$$

Based on (11.13) and (11.22), one verifies that condition $F \in \mathcal{F}(N_1, N_2)$ can be replaced by requirement (1.20) with $R \geq l + (C+1)\Delta$.

Let us discuss the above result briefly. As it follows from (11.28), the nature of iterations (11.8) is essentially different as compared to processes considered before in case of noise in the data or perturbations in the source-type condition. It is for the sake of getting a reasonable approximation, one has to stop those processes at a certain step with the number of this step being a function of the corresponding errors. Unlike these methods, iterations x_n, generated by algorithm (11.8), stabilize as $n \to \infty$ in a neighborhood $B(x^*, (C+1)\Delta)$ of the solution x^*. Therefore method (11.8) does not require a stopping rule.

Comparing (11.27) and (11.28), one can see that estimate (11.28) is most useful when the value of Δ turns out to be considerably less than the acceptable upper bound $(1-q)/(6\gamma N_1 N_2 C)$. From practical point of view, this means that a rather high quality of the approximation of the solution x^* by a finite dimensional affine subspace M_ξ is required. Our experience, gained from prior use of this method for solving applied problems, indicates that the desired quality of the approximation is usually attained when the solution x^* is sufficiently smooth. In this case, frequently one can manage subspaces M of a small dimension as well as the simplest choice of ξ, $\xi = 0$. The possibility to choose $\xi = 0$ illustrates that the original linear subspace M already approximates the element x^* rather accurately, and no additional shift of this subspace is required.

An important part in the above argument is played by condition (11.16). Let us highlight one simple case when this condition is satisfied automatically. This is the case of equation (1.1) with the operator $F'(\tilde{x})$ being injective on some large set of elements $\tilde{x} \in H_1$. For such an operator, the realization of (11.16) is formally ensured if one assumes a priori that the element $\bar{x}^*$ belongs to this set. We point out that injectivity of the derivative operator is fairly common for inverse problems in partial differential equations.

11.3 The case of noisy data

Let us briefly discuss the case when equation (1.1) is given with some error. The above results can easily be carried over the equation

$$F(x) = f, \quad x \in H_1 \quad (f \in H_2) \tag{11.29}$$

with the operator $F \in \mathcal{F}(N_1, N_2)$. For this equation, process (11.8) takes the form

$$x_0 \in M_\xi, \quad x_{n+1} = x_n - \gamma P_M F'^*(x_n)(F(x_n) - f), \quad n = 0, 1, \ldots. \quad (11.30)$$

Suppose now that the exact right-hand side f of equation (11.29) is unknown and, instead, its approximation $\tilde{f} \in H_2$, such that

$$\|\tilde{f} - f\|_{H_2} \leq \delta, \quad (11.31)$$

is available. Then process (11.30) is to be replaced with the iterations

$$x_0 \in M_\xi, \quad x_{n+1} = x_n - \gamma P_M F'^*(x_n)(F(x_n) - \tilde{f}), \quad n = 0, 1, \ldots. \quad (11.32)$$

Keeping all the assumptions made above, instead of (11.11), one gets the estimate

$$\|x_{n+1} - \bar{x}^*\|_{H_1} \leq \|E_1 - \gamma P_M F'^*(\bar{x}^*)F'(\bar{x}^*)P_M\|_{\mathcal{L}(M,M)} \|x_n - \bar{x}^*\|_{H_1}$$
$$+ \frac{3}{2}\gamma N_1 N_2 \|x_n - \bar{x}^*\|_{H_1}^2 + \gamma N_1 \|F(\bar{x}^*) - \tilde{f}\|_{H_2}. \quad (11.33)$$

Using (11.31), one can estimate the last term in (11.33) by $\gamma N_1(N_1 \Delta + \delta)$. Thus, in place of (11.20), one obtains the following resulting inequality

$$\varepsilon_{n+1} \leq q\varepsilon_n + \frac{3}{2}\gamma N_1 N_2 \varepsilon_n^2 + \gamma N_1(N_1 \Delta + \delta), \quad n = 0, 1, \ldots.$$

Applying, with some obvious changes, our previous argument, one concludes that, under the condition on Δ and δ analogous to (11.27) and under the condition

$$\varepsilon_0 = \|x_0 - \bar{x}^*\|_{H_1} \leq l + C(\Delta + \delta)$$

with constants l, C satisfying the assumptions similar to (11.24), (11.26), the estimate

$$\varepsilon_n = \|x_n - \bar{x}^*\|_{H_1} \leq l\tilde{q}^n + C(\Delta + \delta), \quad n = 1, 2, \ldots, \quad (\tilde{q} \in (q, 1))$$

holds. From this estimate one derives

$$\limsup_{n \to \infty} \|x_n - x^*\|_{H_1} \leq C(\Delta + \delta) + \Delta. \quad (11.34)$$

Inequality (11.34) admits the same interpretation as (11.28). Both results emphasize the stable nature of the corresponding iterations as $n \to \infty$ with respect to noise in the input data, as well as with respect to Δ, which characterizes the quality of the approximation of the solution x^* by the subspace M_ξ. In particular, according to (11.34), the ball $B(x^*, C(\Delta + \delta) + \Delta)$ is an attracting set for the iterative points, generated by process (11.32).

In the above discussion, we draw our attention to the existence of the attracting sets (attractors) in the form of the balls of radius proportional to the errors δ and Δ, centered at the solution, for processes (11.8) and (11.32). A more detailed analysis of asymptotic behavior of these processes indicates that, under nonrestrictive additional assumptions, the above processes converge in the norm of H_1 to the certain points from the corresponding balls. The proof of this fact is based on the strong convexity of the functional φ (see (11.3)) in some neighborhood of the point $P_M x^*$, and on the existence of the point of local minimum of φ in this neighborhood. As it is known, for strongly convex functions the method of gradient descent converges to the point of local minimum at a linear rate, [67].

In this chapter the gradient descent method with a constant step size is used for solving problem (11.3). Keeping in mind the property of the local strong convexity of φ, one can replace the gradient method with any other method of finite dimensional minimization that works effectively when applied to problems of minimization of strongly convex functionals. There are corresponding analogues of the above results for all such methods, for example, the steepest descent method, different versions of conjugate gradients methods, as well as Newton's method.

Problems

11.1. Verify identity (11.5).

11.2. Prove that projection $\bar{x}^*$ of the point x^* onto the affine subspace M_ξ is defined by formula (11.9).

11.3. Derive analogues of conditions (11.24), (11.26) and (11.27) for iterations (11.32).

11.4. Find the second derivative of the functional φ and write Newton's method for problem (11.3).

12

Relative computational efficiency of iteratively regularized methods

12.1 Generalized Gauss–Newton methods

The unified approach to the construction of iteratively regularized procedures offers a large variety of methods appropriate for a specific irregular problem. Under the same a priori assumptions many of these methods have the same asymptotic accuracy for a given initial data. Therefore we have to select one or another among methods having the same asymptotic accuracy. It seems reasonable to base our choice on the amount of computational work required for the use of a particular method. The amount of this work is, in fact, asymptotically different for various algorithms. Below we present estimates for the number of operations needed to implement two most used generalized Gauss–Newton methods (10.38). The first one is iteratively regularized Gauss–Newton method (10.3). For this method, $\Theta(\lambda, \alpha) = \frac{1}{\lambda+\alpha}$. The second method is constructed by using $\Theta(\lambda, \alpha) = \mu \sum_{k=0}^{1/\alpha}(1-\mu\lambda)^k$, where $M(n) = 1/\alpha_n$ is a positive integer. The resulting iterations take the form

$$x_{n+1} = \xi - \left\{\mu \sum_{k=0}^{M(n)} (I - \mu F'^*(x_n) F'(x_n))^k\right\} F'^*(x_n)\{F(x_n) - F'(x_n)(x_n - \xi)\}. \tag{12.1}$$

μ is any real number from the interval $(0, 2/N_1^2)$. Under the source condition

$$\xi - x^* \in (F'^*(x^*)F'(x^*))^{1/2} S, \tag{12.2}$$

processes (10.3) and (12.1) converge to x^* and, for both processes, the identity

$$\|x_n - x^*\|_{H_1} = O(\sqrt{\alpha_n}) \tag{12.3}$$

is fulfilled [13]. In practice, operators F and F' can only be known approximately, therefore we assume that F is given by its δ-approximation F_δ; F_δ is also acting from H_1 to H_2, it satisfies conditions (1.2) and (1.3), and

$$\|F_\delta(x^*)\|_{H_2} \leq \delta,$$
$$\|F'_\delta(x) - F'(x)\|_{H_2} \leq \delta, \quad x \in \Omega(x^*). \tag{12.4}$$

12.1 Generalized Gauss–Newton methods

Suppose that one has to find the values of m_1 functions with m unknowns in order to compute F_δ, while the computation of F'_δ results in finding $m_1 m$ functions with m unknowns. Suppose also that it takes $O(\kappa(m))$ arithmetic operations to calculate one value of any m-variable function. Under these assumptions, it requires $m_1 O(\kappa(m))$ and $m_1 m\, O(\kappa(m))$ arithmetic operations to compute F_δ and F'_δ respectively. To simplify our future presentation we set $m_1 = m$. As it follows from the general convergence theory, iterations (10.38) must be terminated when the regularization parameter α_n reaches its critical value that depends on the noise level δ in (12.4). The corresponding value of $n = n(\delta)$ is called a stopping rule. In particular, if the stopping rule for process (10.3) satisfies the condition

$$\alpha_n = \delta, \qquad (12.5)$$

or, more precisely, $n = n(\delta)$ is the integer nearest to the root of (12.5), then

$$\|x_{n(\delta)} - x^*\|_{H_1} = O(\sqrt{\delta}). \qquad (12.6)$$

The same holds for process (12.1), if one assumes that $\alpha_n = 1/M(n)$.

Let us now determine the number of arithmetic operations (AO) needed to implement algorithms (10.3) and (12.1) numerically. It is possible that the number of AO at every step of (10.3) depends on n, but, as it follows from our assumptions, this does not affect the order of the number of operations. Therefore we neglect this dependence. Denote the number of AO required to calculate $F_\delta(x_n)$ and $F'_\delta(x_n)$ by $\nu_1(m)$. The above argument yields $\nu_1(m) = m^2\, O(\kappa(m))$. For each value of n one has to find

$$F'^*_\delta(x_n)\{F_\delta(x_n) - F'_\delta(x_n)(x_n - \xi)\}. \qquad (12.7)$$

This computation requires $\nu_2(m)$ AO, and, since it is reduced to the multiplication of an $m \times m$ matrix by an $m \times 1$ vector, $\nu_2(m) = O(m^2)$. In order to complete the step, it is necessary to calculate the matrix $F'^*_\delta(x_n) F'_\delta(x_n)$ and to invert the matrix $F'^*_\delta(x_n) F'_\delta(x_n) + \alpha_n I$ (or to solve the corresponding linear system). The computation of $F'^*_\delta(x_n) F'_\delta(x_n)$ results in $\nu_3(m)$ AO with $\nu_3(m)$ being equal to $O(m^3)$. The matrix inversion (or solving the corresponding system of linear equations) requires $\nu_4(m)$ AO. For commonly used methods, $\nu_4(m) = O(m^3)$.

Thus, the total number of AO to be performed at every step of process (10.3) is $\sum_{i=1}^{4} \nu_i(m)$, and $n \sum_{i=1}^{4} \nu_i(m)$ AO are needed for n steps of this method. As opposed to (10.3), the number of AO required for one step of process (12.1) largely depends on n. In fact, one component of this number, $\nu_1 + \nu_2$, is independent of n, while the other component, $\lambda(m, n)$, depends on both parameters. In this component, the AO needed to compute the operator-valued polynomial in the right-hand side of (12.1) are taken into consideration. The direct analysis of the right-hand side and our assumptions on the structure of the approximate operators result in the following estimate

$$\lambda(m, n) = M(n)\, O(m^2).$$

Note, that in case of (12.1) one does not have to form the linear operator $F'^*_\delta(x_n)F'_\delta(x_n)$, since inner iterations can be reduced to matrix-vector multiplications only. The total number of AO needed to execute n steps of process (12.1) is

$$n(v_1 + v_2) + \sum_{i=1}^{n} \lambda(m, i).$$

Clearly, the estimates for the number of AO required to implement methods (10.3) and (12.1) have the term $n(v_1 + v_2)$ in common, which can, therefore, be neglected upon comparative analysis. The remaining terms of the estimates have the following asymptotic behavior

$$n(v_3 + v_4) = n\, O(m^3) \quad \text{for method (10.3) and}$$

$$\sum_{i=1}^{n} M(i)\, O(m^2) \quad \text{for method (12.1).}$$

Let

$$M(i) = 2^i. \tag{12.8}$$

Under this assumption, it follows from (12.5) that $n(\delta) \approx \ln(1/\delta)$ and $\sum_{i=1}^{n(\delta)} M(i) \approx O(1/\delta)$, and the comparable parts of the estimates are of order $\ln(1/\delta) O(m^3)$ and $(1/\delta)\, O(m^2)$ for methods (10.3) and (12.1) respectively. Hence, method (12.1) is preferable to iteratively regularized Gauss–Newton method (10.3) only if

$$m \gg 1/(|\ln \delta|\delta). \tag{12.9}$$

For example, if $\delta = 10^{-4}$, then the advantage of method (12.1) (as compared to (10.3)) may become evident when m is much larger than 1000. For smaller values of m, process (12.1) can still be superior to process (10.3) if a priori conditions on $\xi - x^*$ are more restrictive than (12.2).

12.2 A more restrictive source condition

It is shown in [13], that under a priori condition

$$\xi - x^* \in (F'^*(x^*)F'(x^*))^k S, \quad k \geq 1, \tag{12.10}$$

the best stopping rule $n = n(\delta)$ for method (10.3), regardless of k, is the one that satisfies the identity $\alpha_n \approx \delta^{2/3}$. In particular, if $\alpha_n = 2^{-n}$, then $n(\delta) \approx \ln(1/\delta)$ and

$$\|x_{n(\delta)} - x^*\|_{H_1} = O(\delta^{2/3}). \tag{12.11}$$

On the other hand, the convergence rate of method (12.1) improves as the exponent k gets bigger. The optimal stopping rule for $\alpha_n = 2^{-n}$ is $n(\delta) \approx (1/k) \ln(1/\delta)$, and

$$\|x_{n(\delta)} - x^*\|_{H_1} = O(\delta^{\frac{2k}{2k+1}}). \tag{12.12}$$

Suppose condition (12.10) is fulfilled, and the best stopping rules for processes (10.3) and (12.1) are used. Let us compute the asymptotic numbers of AO for the two processes. Denote these numbers L_1 and L_2 respectively. The argument is similar to the one applied in case of condition (12.2). One has

$$L_1 = \ln(1/\delta)\, O(m^2 \kappa(m)) = n(\delta) \sum_{i=1}^{4} v_i(m), \qquad (12.13)$$

and

$$L_2 = n(\delta)(v_1 + v_2) + \sum_{i=1}^{n(\delta)} \lambda(m,i) = \frac{1}{k}\ln(1/\delta)(v_1 + v_2) + \frac{1}{\delta^{1/k}} O(m^2)$$

$$= \frac{1}{k}\ln(1/\delta)\, O(m^2 \kappa(m)) + \frac{1}{\delta^{1/k}} O(m^2). \qquad (12.14)$$

One can see from (12.13) and (12.14) that L_1 does not depend on the exponent k in (12.10), while L_2 is a decreasing function of k. Thus, for any value of m, under the reasonable assumption

$$m/\kappa(m) = O(1), \qquad (12.15)$$

method (12.1) is preferable to method (10.3) in terms of AO.

12.3 Comparison to iteratively regularized gradient scheme

In this section we calculate the number of AO needed for the implementation of iteratively regularized gradient scheme (9.5). This method has an attractively simple structure. The iterations are performed by the following rule

$$x_{n+1} = x_n - \gamma_n \{F'^*(x_n) F(x_n) + \alpha_n (x_n - \xi)\}, \qquad \gamma_n, \alpha_n > 0. \qquad (12.16)$$

The reader may consult [13, 15] for the convergence analysis of (12.16) in case of a general smooth operator F. The sequence $\{\gamma_n\}$ in (12.16) depends on N_1 and N_2. For further inferences, the values of γ_n are irrelevant. Under a priori condition (12.2) and stopping criteria (12.5), identity (12.6) holds as well. Unfortunately, in case of (12.16) one cannot manipulate the sequence α_n like in processes (10.3) and (12.1). In particular, one cannot use $\alpha_n = 2^{-n}$. Relationship (12.6) is guaranteed only if α_n is decreasing "slowly" [13]. Presently known results ensure that (12.6) is satisfied when α_n is decreasing not faster than $1/\sqrt{n}$. If we set $\alpha_n = 1/\sqrt{n}$ and choose the stopping rule according to (12.5), then $n(\delta) \approx 1/\delta^2$. Since at every step the number of AO is $v_1 + v_2 = O(m^2 \kappa(m))$,

$$L_3 = \frac{1}{\delta^2} O(m^2 \kappa(m)) \qquad (12.17)$$

AO are required for the necessary number of iterations (12.16). Comparing processes (10.3), (12.1) and (12.16) one can see that, at least for $\delta \ll 1$, implicit methods (10.3) and (12.1) are superior to explicit method (12.16).

Thus, under the least restrictive a priori assumptions (12.2) and for sufficiently small m and δ, process (10.3), which is seemingly the most difficult, turns out to be the most efficient in terms of accuracy and asymptotic number of AO.

Problems

12.1. Find the exact number arithmetic operations required to multiply two $m \times m$ matrices.

12.2. Show that the total number of multiplications and divisions required to solve a nonsingular system of m linear equations with m unknowns by LU factorization is $\frac{m^3}{3} + m^2 - \frac{m}{3} \approx \frac{m^3}{3}$ [8].

12.3. Show that the total number of additions and subtractions required to solve a nonsingular system of m linear equations with m unknowns by LU factorization is $\frac{m(m-1)(2m+5)}{6} \approx \frac{m^3}{3}$ [8].

12.4. Suppose the determinants in Cramer's rule for solving $m \times m$ nonsingular linear system are computed using expansion by minors. Prove that the number of multiplications and divisions required to implement Cramer's rule is $(m+1)!$ [8]. How does it compare to LU factorization if $m = 10$?

12.5. Verify that the number of multiplications and divisions required to evaluate A^{-1} is $\frac{4m^3}{3} - \frac{m}{3} \approx \frac{4m^3}{3}$ [8], which shows that calculating A^{-1} is only four times the expense of solving $Ax = b$ for a single vector b, not m times the expense as one might intuitively assume.

13

Numerical investigation of two-dimensional inverse gravimetry problem

13.1 Problem formulation

To test the numerical efficiency of iteratively regularized gradient method (9.5) presented in Chapter 9, the two-dimensional inverse gravimetry problem was considered. It was assumed that the lower half-space is formed by two media with constant densities separated by a surface $x = x(\varsigma, \nu)$ [2]. The gravitational anomaly, $f = f(t, s)$, is caused by the deviation of the above surface from the horizontal plane $x(\varsigma, \nu) = h$. It is well known that under the above assumptions in the Descartes coordinate system the *inverse gravimetry problem* can be reduced to the 2D nonlinear integral equation of the first kind:

$$F(x) := g \, \Delta\sigma \int_a^b \int_c^d \left\{ \frac{1}{[(\varsigma - t)^2 + (\nu - s)^2 + x^2(\varsigma, \nu)]^{1/2}} \right.$$
$$\left. - \frac{1}{[(\varsigma - t)^2 + (\nu - s)^2 + h^2]^{1/2}} \right\} d\varsigma \, d\nu = f_\delta(t, s), \tag{13.1}$$

and it consists of finding the unknown function $x = x(\varsigma, \nu)$, which describes the interface, from the measured data $f_\delta = f_\delta(t, s)$. Here g is the gravitational constant and $\Delta\sigma$ is the density jump on the interface.

Take $H_1 = H^1(\Omega)$ and $H_2 = L_2(\Omega)$, where $\Omega = [a, b] \times [c, d]$ and $H^1(\Omega) = W^{1,2}(\Omega)$ with $W^{k,p}(\Omega)$ being defined as the subset of L_p such that a function φ and its derivatives up to some order k have a finite L_p norm, for given $p \geq 1$. If one denotes the kernel of the nonlinear operator $F(x)$ by $K(t, s, \varsigma, \nu, x(\varsigma, \nu))$:

$$K(t, s, \varsigma, \nu, x(\varsigma, \nu)) := \left\{ \frac{1}{[(\varsigma - t)^2 + (\nu - s)^2 + x^2(\varsigma, \nu)]^{1/2}} \right.$$
$$\left. - \frac{1}{[(\varsigma - t)^2 + (\nu - s)^2 + h^2]^{1/2}} \right\}, \tag{13.2}$$

then the Fréchet derivative of $F(x)$ is given by

$$F'(x)y = g \, \Delta\sigma \int_\Omega \int K'_x(t, s, \varsigma, \nu, x(\varsigma, \nu)) y(\varsigma, \nu) \, d\varsigma \, d\nu, \qquad (13.3)$$

where

$$K'_x(t, s, \varsigma, \nu, x(\varsigma, \nu)) = -\frac{x(\varsigma, \nu)}{[(\varsigma - t)^2 + (\nu - s)^2 + x^2(\varsigma, \nu)]^{3/2}}. \qquad (13.4)$$

To evaluate the corresponding adjoint operator, one needs to solve an auxiliary integro-differential equation. Hence direct numerical realization of the iteratively regularized gradient method is very complicated in our case. It was shown in [10] that one can find an approximate solution to (13.1) by replacing the integral in (13.1) with a quadrature formula. Thus from now on, we consider a finite-dimensional analogue of equation (13.1), and construct iterations (13.6) in a suitable Euclidean space. In a Euclidean space the corresponding adjoint operator $F'^*(x)$ is a discrete analogue of the following linear integral operator

$$\mathcal{A}z := g \, \Delta\sigma \int_\Omega \int K'_x(t, s, \varsigma, \nu, x(\varsigma, \nu)) z(t, s) \, dt \, ds. \qquad (13.5)$$

13.2 The algorithm

For equation (13.1) and $\xi = x_0$, iteratively regularized gradient method (9.5) takes the following form

$$x_0 \in H_1, \quad x_{n+1} = x_n - \gamma_n \{F'^*(x_n)(F(x_n) - f_\delta) + \alpha_n(x_n - x_0)\}. \qquad (13.6)$$

Here γ_n is an a priori prescribed step size, and α_n is a regularization parameter at the nth iteration.

Suppose f_δ approximates the exact data f with the accuracy δ, i.e.,

$$\|f_\delta - f\|_{H_2} \leq \delta. \qquad (13.7)$$

In the case when $\delta > 0$, the sequence $\{x_n\}$ generated by iteratively regularized process (13.6) does not usually converge to a solution $\hat{x}$ of (13.1). It does, however, allow a stable approximation to $\hat{x}$ if iterations are stopped at an appropriate step $n = N(\delta)$ such that $\lim_{\delta \to 0} \|x_{N(\delta)} - \hat{x}\| = 0$. For the experiment below, the stopping number $n = N(\delta)$ was chosen by a *generalized discrepancy principle*, i.e., iterations (13.6) were stopped at the first index N, for which the residual $\|F(x_N) - f_\delta\|$ is less than $\sqrt{\tau\delta}$, $\tau > 1$:

$$\|F(x_N) - f_\delta\| < \sqrt{\tau\delta} \leq \|F(x_n) - f_\delta\|, \quad 0 \leq n < N, \quad \tau > 1. \qquad (13.8)$$

We refer to [15] for a rigorous convergence analysis of algorithm (13.6)–(13.8).

13.3 Simulations

The numerical experiments presented in this chapter were performed by the authors in collaboration with M. A. Skinner, who, at the time, was a graduate student at the Department of Mathematics and Statistics, Georgia State University. The programming was done in Matlab 6.5.1 on a Pentium IV desktop computer. Discretization was implemented by taking an evenly spaced grid over the domain $\Omega = [2.8, 20.0] \times [0.0, 8.0]$ (km^2) with 43×21 node points, respectively. This resolution gave the corresponding mesh width of $w = 0.4$(km) in both directions. The two-dimensional analogue of the midpoint quadrature rule was used to approximate the integral operator in (3.1). The ground surface height was taken to be $h = 2.0$(km). The gravitational anomaly was calculated by solving the direct problem with $x = x_{\text{mod}}(\varsigma, \nu)$, and the identity was rescaled by dropping the constants g and $\Delta \sigma$ from the original formula. The constant horizontal plane $x_0(\varsigma, \nu) = 0.1$(km) was taken as the initial guess for all the simulations below.

In order to select the regularization parameter α_n, the step size parameter γ_n, and the discrepancy parameter τ, the following representative, asymmetric model solution, obtained by translating and scaling Gaussian distributions, was considered

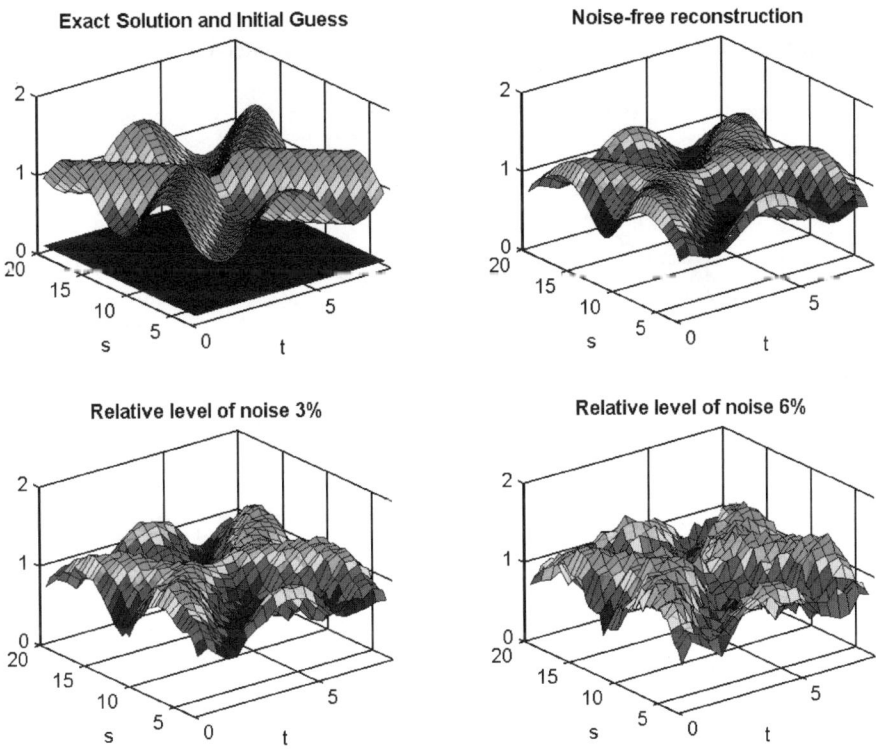

Fig. 13.1.

(see 'Built-in Peaks Surface Function' in Matlab manual):

$$x_{\text{mod}}^{(1)}(\tilde{\varsigma}, \tilde{v}) = \frac{1}{14} \left\{ 3(1-\tilde{\varsigma})^2 e^{-\tilde{\varsigma}^2 - (\tilde{v}+1)^2} \right.$$
$$\left. -10 \left(\frac{1}{5}\tilde{\varsigma} - \tilde{\varsigma}^3 - \tilde{v}^5 \right) e^{-\tilde{\varsigma}^2 - \tilde{v}^2} - \frac{1}{3} e^{-(\tilde{\varsigma}+1)^2 - \tilde{v}^2} \right\} + 1, \quad (13.9)$$

where

$$\tilde{\varsigma} = 6 \left(\frac{\varsigma - c}{d - c} \right) - 3, \quad \text{and} \quad \tilde{v} = 6 \left(\frac{v - a}{b - a} \right) - 3$$

are appropriate domain transformations. For the above model solution the sequence γ_n was numerically determined so as to give the most aggressive convergence rate for the experiments conducted in a noise-free case and in the presence of noise at the level from 2% to 6%. The best convergence rate was obtained for $\gamma_n = \gamma_0 = 0.015$. The regularization parameter sequence $\alpha_n = \alpha_0 (1+n)^{-0.25}$ with $\alpha_0 = 0.001$ was chosen to ensure the best stability in the presence of noise. For the noise-free case there was no apparent need for regularization (we could use $\alpha_n = 0$). An attractive feature of scheme (13.6) is that it is not very sensitive to the values of parameters $\{\gamma_n\}$ and $\{\alpha_n\}$. For the relative noise 6% we were able to take α_0 from the interval $[10^{-6}, 10^{-2}]$ and $\alpha_n = \frac{\alpha_0}{(1+n)^p}$ with $0 < p \le 1$. One can see that for our particular problem the interval for p is bigger than the interval guaranteed by the convergence

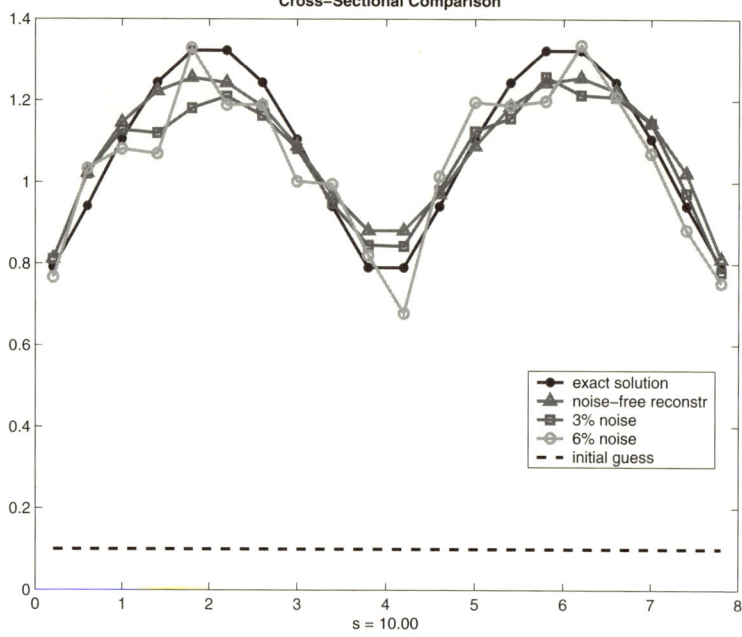

Fig. 13.2.

13.3 Simulations

theorem. As for the choice of γ_n, if one takes $\gamma_n = \gamma_0$ then one can use any $\gamma_0 \in [0.005, 0.03]$. The sequence $\gamma_n = \frac{\gamma_0}{(1+n)^q}$, $0 < q \le 1$ and $\gamma_0 \in [0.005, 0.03]$, also works. For $q > 1$ convergence becomes rather slow.

The following algorithm was used to determine τ. Iterations (13.6) were performed with random noise functions added to the right-hand side of equation (13.1). The relative discrepancy:

$$\Delta_\delta := \frac{\|F(x_{\mathrm{mod}}) - f_\delta\|}{\|F(x_{\mathrm{mod}})\|} \qquad (13.10)$$

ranged from 0.02 to 0.2. All together 10 different noise functions were investigated with $\Delta_{\delta(k)} = 0.02k$, $k = 1, 2, \ldots, 10$. For every noise function, iterative process (13.6) was stopped at the first number $n = n(k)$ such that

$$\frac{\|x_n - x_{\mathrm{mod}}\|}{\|x_{\mathrm{mod}}\|} > \frac{\|x_{n-1} - x_{\mathrm{mod}}\|}{\|x_{\mathrm{mod}}\|}$$

and $\tau = \tau(k)$ was calculated from the identity

$$\|F(x_n) - f_\delta\|^2 = \tau \delta.$$

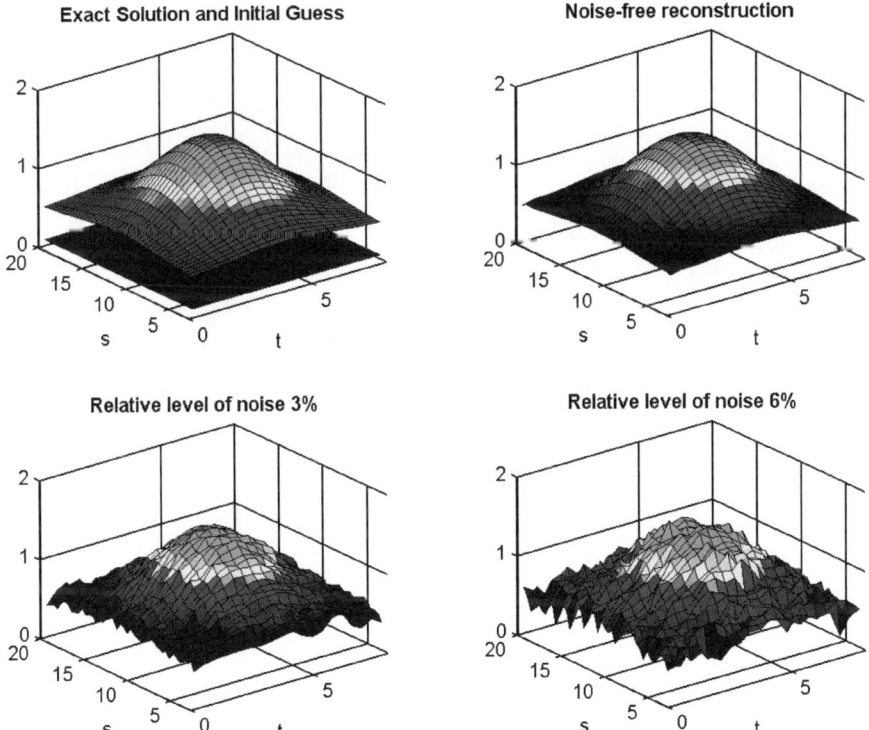

Fig. 13.3.

It was discovered that for $k = 1, 2, \ldots, 10$, $\tau(k) \in [8.02, 11.46]$. As the result, the value $\tau = 11.46$ was taken to reconstruct two other model solutions

$$x_{\text{mod}}^{(2)}(\tilde{\varsigma}, \tilde{\nu}) = \sin(|\tilde{\varsigma}| - |\tilde{\nu}|)/3 + 1, \quad \tilde{\varsigma} = 10\left(\frac{\varsigma - c}{d - c}\right) - 5, \quad \tilde{\nu} = 10\left(\frac{\nu - a}{b - a}\right) - 5$$

and

$$x_{\text{mod}}^{(3)}(\tilde{\varsigma}, \tilde{\nu}) = \exp(-\tilde{\varsigma}^2 - \tilde{\nu}^2) + 0.5, \quad \tilde{\varsigma} = 3\left(\frac{\varsigma - c}{d - c}\right) - 1.5, \quad \tilde{\nu} = 3\left(\frac{\nu - a}{b - a}\right) - 1.5$$

using method (13.6) and a posteriori stopping rule (13.8) (with parameters $\gamma_n = 0.015$ and $\alpha_n = 0.001(1 + n)^{-0.25}$).

In Figure 13.1, one can see the graph of $x_{\text{mod}}^{(2)}(t, s)$, $(t, s) \in \Omega$, as well as the graphs of approximate solutions for the noise-free case, for the case when $\Delta_\delta := 0.02$ and $\Delta_\delta := 0.05$ (see formula (13.10)). The cross-sectional comparison for $s = 10$ is presented in Figure 13.2. The same results for $x_{\text{mod}}^{(3)}(t, s)$ are illustrated in Figures 13.3 and 13.4, respectively. The number of iterations in the noise-free case for both model solutions was 50. When the relative level of noise was 2% and the experiment was conducted with $x_{\text{mod}}^{(3)}(t, s)$, the iterations were stopped by generalized discrepancy principle (13.8) at $n = 31$. When the level of noise was 6%, for the same model solution, the iterations were stopped at $n = 21$.

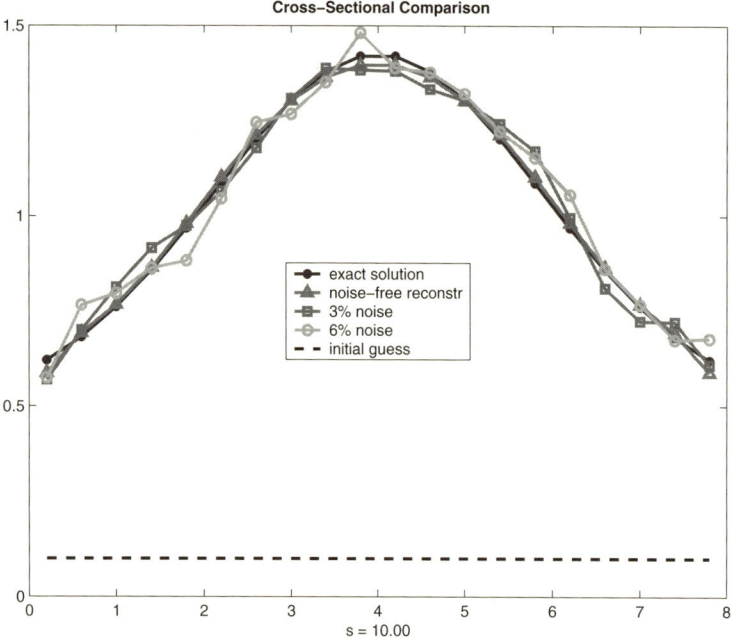

Fig. 13.4.

Problems

13.1. Consider 1D inverse gravimetry problem [77] described by the following nonlinear operator equation

$$F(x) := \frac{\rho}{4\pi} \int_{-l}^{l} K(t, s, x(s)) ds = f(t), \qquad (13.11)$$

where

$$K(t, s, x(s)) = \ln \frac{(t-s)^2 + h^2}{(t-s)^2 + (h - x(s))^2}. \qquad (13.12)$$

It is assumed here that the sources of a gravitational field with a constant density ρ are distributed in the domain

$$D = \{-l \leq t \leq l, \; -h \leq z \leq -h + x(s)\},$$

where $x(s)$ is an interface between two media (with and without the sources), $f(t)$ is the gravity strength anomaly, l and h are parameters of the domain. The operator F acts between the pair of Hilbert spaces $H_1 = H^1[-l, l]$ and $H_2 = L_2[-l, l]$. Calculate the Fréchet derivative $F'(x)$ of the operator F.

13.2. Solve 1D forward gravimetry problem, i.e., determine the function $f(t)$ for the model solution $x_{\text{mod}}(s) = (1 - s^2)^2$ by calculating the integral in the left-hand side of (13.11) with a quadrature formula. Take $l = 1, h = 2, \rho = 1$.

13.3. Now given the function $f(t)$ computed in the previous exercise, solve 1D inverse gravimetry problem, that is, reconstruct $x_{\text{mod}}(s)$ numerically using iterative method (13.6). Replace $f(t)$ with its δ-approximation $f_\delta(t)$, such that $\|f_\delta - f\|_{H_2} \leq \delta$, by adding random noise to the function $f(t)$. Reconstruct $x_{\text{mod}}(s)$ numerically using algorithm (13.6)–(13.8). Take the relative level of noise to be 5%, and the initial approximation $x_0(s) = 1$.

13.4. Repeat Exercise 13.3 applying iteratively regularized Gauss–Newton (IRGN) method

$$x_{n+1} = \xi - (F'^*(x_n) F'(x_n) + \alpha_n E_1)^{-1} F'^*(x_n) [F(x_n) - f_\delta - F'(x_n)(x_n - \xi)],$$

investigated in Chapter 10, in place of iteratively regularized gradient scheme (13.6).

13.5. Consider the *inverse scattering problem* of determining the shape of an obstacle from far-field scattering data [79]. Under the assumption that the scattering region Ω is starlike about the origin, i.e., the boundary of Ω can be parameterized by a continuous polar curve $R(\theta)$, $0 \leq \theta < 2\pi$, the problem can be reduced to the nonlinear integral equation of the first kind:

$$d(\psi) = \int_0^{2\pi} g(\theta, \psi, R(\theta)) \, d\theta. \qquad (13.13)$$

The kernel g is the following expression

$$g(\theta, \psi, r) := [\exp(bR)(bR - 1) + 1]/b^2, \quad b := -2i k_0 \cos(\psi - \theta). \quad (13.14)$$

The angle ψ gives the direction of the incident field with respect to the positive x-axis, and k_0 is the wavenumber for the wavefield in the host medium.

Calculate the scattering data by solving the direct problem with a peanut-shape curve [79]

$$R = R_{\text{exact}}(\theta) = \left(\cos^2 \left(\theta - \frac{\pi}{4} \right) + 0.25 \sin^2 \left(\theta - \frac{\pi}{4} \right) \right)^{1/2}, \quad (13.15)$$

$0 \leq \theta \leq 2\pi$. Graph the real and imaginary parts of the computed scattering data $d(\psi)$ for incident angle ψ between 0 and $2\pi/3$.

13.6. Define the nonlinear operator $F : \mathbb{R}^k \to \mathbb{C}^l$ by

$$(F(\mathbf{R}))_m := \sum_{j=1}^{k} \mu_j g(\theta_j, \psi_m, R_j) - d_m, \quad m = 1, \ldots, l, \quad (13.16)$$

where μ_j are the coefficients of a quadrature formula, $\mathbf{d} = [d_1, \ldots, d_l]^T$ is the vector of scattering data points, and $\mathbf{R} = [R_1, \ldots, R_k]$ is an approximate solution. The Fréchet derivative (Jacobian) of this operator is the $l \times k$ matrix $F'(\mathbf{R})$ with entries

$$(F'(\mathbf{R}))_{mj} := \mu_j \partial g / \partial R(\theta_j, \psi_m, R_j), \quad m = 1, \ldots, l, \quad j = 1, \ldots, k. \quad (13.17)$$

Given the scattering data obtained in Exercise 13.5, solve the inverse problem numerically (i.e., reconstruct the discrete analogue of $R_{\text{exact}}(\theta)$) by IRGN algorithm

$$\mathbf{R}_{n+1} = \xi - (F'^*(\mathbf{R}_n) F'(\mathbf{R}_n) + \alpha_n E_1)^{-1} F'^*(\mathbf{R}_n)[F(\mathbf{R}_n) - \mathbf{d} - F'(\mathbf{R}_n)(\mathbf{R}_n - \xi)],$$

Use $\mathbf{R}_0(\theta) = 1$ as the initial guess. Graph both exact and approximate solutions in the same polar coordinate system.

13.7. Perform numerical simulations described in this chapter for 2D *inverse magnetometery problem* [2]

$$F(x) := \Delta J \int_a^b \int_c^d \left\{ \frac{x(\varsigma, \nu)}{[(\varsigma - t)^2 + (\nu - s)^2 + x^2(\varsigma, \nu)]^{3/2}} \right.$$

$$\left. - \frac{h}{[(\varsigma - t)^2 + (\nu - s)^2 + h^2]^{3/2}} \right\} d\varsigma \, d\nu = y_\delta(t, s).$$

$$(13.18)$$

It consists of finding the unknown function $x = x(\varsigma, \nu)$, which describes the interface, from the measured anomalous magnetic field $y_\delta(t, s)$. In (13.18), ΔJ is the averaged jump of the vertical component of the magnetization vector.

14

Iteratively regularized methods for inverse problem in optical tomography

14.1 Statement of the problem

In this section the numerical efficiency of iteratively regularized methods is illustrated by simulations for a two dimensional version of the ill-posed optical tomography inverse problem. In optical imaging, low-energy visible light (wavelength from 700 to 1200 nm) is used to illuminate the biological tissue. The illumination of the tissue can be modeled as a photon transport phenomenon. The process is described by the most widely applied equation in optical imaging, the *radiative transfer or transport equation* (RTE) [7]. A low order *Diffusion Approximation* (DA) to the RTE

$$\frac{1}{c}\frac{\partial u}{\partial t}(\mathbf{x},t) - \nabla \cdot D(\mathbf{x})\nabla u(\mathbf{x},t) + \mu_a(\mathbf{x})u(\mathbf{x},t) = f(\mathbf{x},t) \tag{14.1}$$

together with initial condition,

$$u(\mathbf{x},0) = 0 \quad \text{in } \Omega \in \mathbb{R}^n \quad (n = 1, 2 \text{ or } 3) \tag{14.2}$$

and the boundary condition,

$$u(\mathbf{x},t) + 2D(\mathbf{x})\frac{\partial u}{\partial \nu}(\mathbf{x},t) = 0, \quad \mathbf{x} \in \partial\Omega, \tag{14.3}$$

has been derived and studied in the last several years, [7]. Here $\mathbf{x}$ is the spatial variable on the domain Ω, ν is the normal direction from the boundary and f represents the forcing function, or source. The two parameters, the coefficient of diffusion D and the coefficient of absorption μ_a, characterize the internal optical properties of the media.

In the time independent case the diffusion approximation can be written as follows

$$-\nabla \cdot D(\mathbf{x})\nabla u(\mathbf{x}) + \mu_a(\mathbf{x})u(\mathbf{x}) = f(\mathbf{x}). \tag{14.4}$$

The associated boundary condition is

$$u(\mathbf{x}) + 2D(\mathbf{x})\frac{\partial u}{\partial \nu}(\mathbf{x}) = 0, \quad \mathbf{x} \in \partial\Omega. \tag{14.5}$$

The forward problem can be defined as: given sources $f = f_j$, $j = 1, 2, \ldots, n_s$ in Ω and $\mathbf{q}$ in Q, a vector of model parameters (the coefficient of diffusion D and the coefficient of absorption μ_a, i.e., $\mathbf{q} = (D, \mu_a)^T$), find the data u on $\partial\Omega$. The *inverse problem* is: given sources f_j in Ω and data u on $\partial\Omega$, find $\mathbf{q}$ in Q.

This DA has been widely used to calculate photon migration in biological tissues [38], and the existing computational methods for the inverse problem are almost exclusively based on the DA [24].

In general, measurement of $u(\mathbf{q})$ may not be possible, only some observable part $\mathcal{C}u(\mathbf{q})$ of the actual state $u(\mathbf{q})$ may be measured. In this abstract setting, the objective of the inverse parameter estimation problem is to choose a parameter $\mathbf{q}^*$ in Q, that minimizes an error criterion or cost functional $\Phi(u(\mathbf{q}), \mathcal{C}u(\mathbf{q}), \mathbf{q})$ over all possible $\mathbf{q}$ in Q subject to $u(\mathbf{q})$ satisfying the diffusion approximation. A typical observation operator is

$$\begin{aligned}\mathcal{C}u(\mathbf{q}) &= \left\{-D\frac{\partial u}{\partial \nu}(\mathbf{x}_i; \mathbf{q}, f_j)\right\}_{i=\overline{1,n_m}, j=\overline{1,n_s}} \\ &= \left\{\frac{1}{2}u(\mathbf{x}_i; \mathbf{q}, f_j)\right\}_{i=\overline{1,n_m}, j=\overline{1,n_s}},\end{aligned} \tag{14.6}$$

where $\mathbf{x}_i$ is in $\partial\Omega$, n_m is the number of measurements, n_s is the number of sources, and the second equality comes from boundary condition (14.5). Thus we get the following *nonlinear least square problem*

$$\min_{\mathbf{q}} J(\mathbf{q}) = \min_{\mathbf{q}} \sum_{j=1}^{m_s}\sum_{i=1}^{m} |\mathcal{C}_{ij}u(\mathbf{q}) - z_{ij}|^2 = \min_{\mathbf{q}} \|\mathbf{U}(\mathbf{q}) - \mathbf{z}_\delta\|^2. \tag{14.7}$$

Here z_{ij} is the measured data at the boundary for a given source f_j, and $\|\mathbf{z}_\delta - \mathbf{z}\| \leq \delta$, $\delta > 0$ is the accuracy of measurements.

14.2 Simple example

In what follows, we illustrate the theoretical concepts with a simple example in one dimension [49]. Let $\Omega = [0, L]$, the diffusion approximation with constant background is the Sturm–Liouville equation:

$$-u'' + \mathbf{q}^2 u = \frac{f}{D} \tag{14.8}$$

where $\mathbf{q}^2 = \mu_a/D$ is constant, with the Rubin boundary condition:

$$\begin{aligned}u(0) - 2Du'(0) &= 0 \\ u(L) + 2Du'(L) &= 0.\end{aligned} \tag{14.9}$$

The inverse problem is to estimate the scalar $\mathbf{q}$ from the data $\mathbf{z}$ measured at $\mathbf{x} = 0$ or $\mathbf{x} = L$.

For example, Green's function (solution for a delta distribution source $f(\mathbf{x}) = \delta(\mathbf{x} - \mathbf{x}_s)$ where $\mathbf{x}_s$ is the location of the source) for problem (14.8)–(14.9) can be computed analytically and the solution for $\mathbf{x} < \mathbf{x}_s$ is:

$$u(\mathbf{x}, \mathbf{x}_s; \mathbf{q}) = \frac{(e^{\mathbf{q}\mathbf{x}_s} - \gamma e^{-\mathbf{q}\mathbf{x}_s})(e^{\mathbf{q}\mathbf{x}} - \beta e^{-\mathbf{q}\mathbf{x}})}{2\mathbf{q} D(\beta - \gamma)} \tag{14.10}$$

where

$$\gamma = \frac{e^{2L\mathbf{q}}}{\beta}$$

$$\beta = \frac{1 - 2D\mathbf{q}}{1 + 2D\mathbf{q}}.$$

If we measure $-D\nabla u \cdot v$ at $\mathbf{x}_1 = 0$, then the inverse problem is to estimate $\mathbf{q}$ from the parameter to output map given by the formula

$$\begin{aligned} \mathbf{U}(\mathbf{q}) &= \mathcal{C}u(\mathbf{x}_1, \mathbf{x}_s; \mathbf{q}) \\ &= -D\frac{\partial u}{\partial v}(0, \mathbf{x}_s; \mathbf{q}) \\ &= \frac{(e^{\mathbf{q}\mathbf{x}_s} - \gamma e^{-\mathbf{q}\mathbf{x}_s})(1 + \beta)}{2(\beta - \gamma)}, \end{aligned} \tag{14.11}$$

which is a nonlinear function of the parameter $\mathbf{q}$ as discussed. In Figure 14.1, we plot the Tikhonov functional

$$J_\lambda(\mathbf{q}) := |\mathbf{U}(\mathbf{q}) - \mathbf{z}_1|^2 + \lambda |\mathbf{q}|^2$$

for the 1D diffusion approximation with Rubin boundary conditions for a homogeneous background medium with $\mu_a = 0.012$ mm^{-1} and $D = 0.33$ mm.

This is the simulation of the 1D diffusion approximation on the interval $(0, 43.0)$ with $\mathbf{q} = \sqrt{\mu_a/D}$. We computed the cost functional $J_\lambda(\mathbf{q})$ for $\mathbf{q}^* = 0.1907$ mm^{-1} (corresponding to $\mathbf{z}_1 = \mathbf{U}(\mathbf{q}^*)$, $\mathbf{x}_1 = 0$, $\mu_a = 0.012$ mm^{-1}, and $D = 0.33$ mm) over a range of $\mathbf{q}$ starting from 0.14 to 0.4. The solid curve represents J_λ without regularization ($\lambda = 0$) and the broken curve represents J_λ with regularization parameter $\lambda = 10^{-5}$. From Figure 14.1, it is clear that without regularization ($\lambda = 0$) the functional J_λ is rather insensitive to the parameter $\mathbf{q} = \sqrt{\mu_a/D}$ (i.e. the numerical method starting with an overestimate of the true parameter is bound to fail). But with regularization ($\lambda = 10^{-5}$), J_λ is convex. We note here that the regularization has changed the problem so that we are solving for a minimum $\mathbf{q}_\lambda$ that is no longer the same as $\mathbf{q}^*$, the solution to our original problem.

This simple example for a constant background illustrates the complexity of a nonlinear ill-posed inverse problem. In the rest of the chapter we investigate further

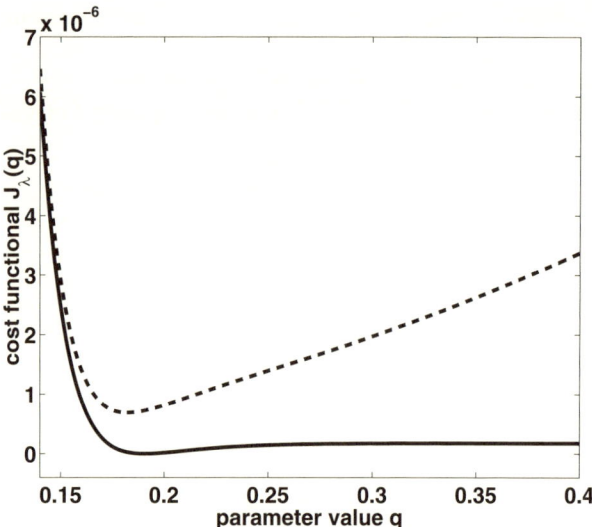

Fig. 14.1. Parameter estimation and regularization.

the two-dimensional inverse problem in general with spatially varying parameters D and μ_a.

14.3 Forward simulation

Solving the forward problem amounts to solving a PDE. The general analytic method for solving PDEs containing a delta distribution source is the Green's function method. However for complex geometries, the analytic solution is intractable. Therefore one requires numerical solutions. The *finite element method* (FEM) is more versatile than other methods including the finite difference method because of its ease in complex geometries and modeling boundary effects. The FEM is a variational method used to approximate the solution by a family of finite dimensional basis functions. Then the forward problem is reduced to one of linear algebra. Here we illustrate the use of the FEM for solving 2D time-independent forward problem following our joint paper with N. Tuncer (see [17]).

The diffusion approximation of RTE is an elliptic partial differential equation with Robin boundary conditions [26, 20, 21]. The weak formulation of forward problem (14.4)–(14.5) is to find $u \in H^1(\Omega)$ such that for any $v \in H^1(\Omega)$

$$\int_\Omega \nabla v \cdot D \nabla u \, d\varsigma \, dv + \int_\Omega v \mu_a u \, d\varsigma \, dv + \frac{1}{2} \int_{\partial\Omega} vu \, ds = \int_\Omega v f \, d\varsigma \, dv. \quad (14.12)$$

Galerkin approximation to the solution of the above weak problem (14.12) is simply based on constructing the discrete analogue of (14.12) on a finite dimensional subspace of $H^1(\Omega)$ [26, 21].

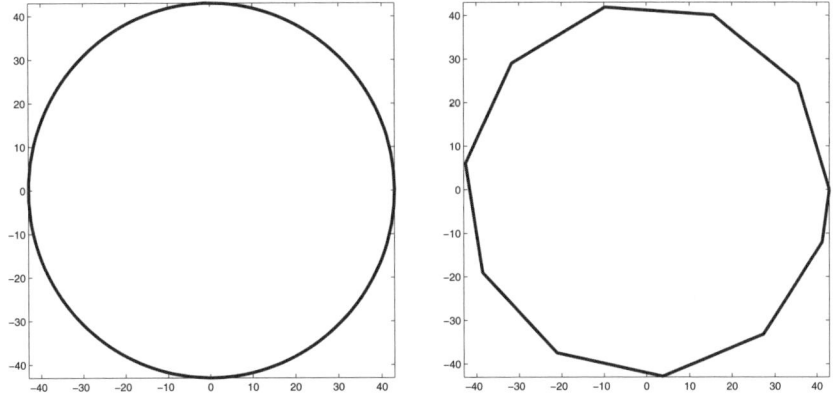

Fig. 14.2. The boundary and its linear approximation.

We define a finite dimensional subspace of $H^1(\Omega)$, by first partitioning the domain Ω. It is very common in the finite element theory to use a linear approximation of the boundary in curved domains such as Ω. In optical tomography we assume Ω to be a ball in $\mathbb{R}^n$ with radius r, i.e.,

$$\Omega = \{\mathbf{x} \in \mathbb{R}^n : \|\mathbf{x}\|_2 \leq r\}.$$

Let Ω_h be a linear approximation of Ω, and let $\mathcal{K}_1, \mathcal{K}_2, \ldots, \mathcal{K}_n$ be triangles or quadrilaterals such that they form a partition of Ω_h, i.e

$$\Omega_h = \bigcup_{i=1}^{n} \mathcal{K}_i.$$

For any intersection of such two triangles or quadrilaterals $\mathcal{K}_i$ and $\mathcal{K}_j$, if the intersection consists of one point then it is a common vertex, if the intersection consists of more than one point then it is a common edge [20, 26]. Let $\mathcal{T}^j = \{\mathcal{K}_1^j, \mathcal{K}_2^j, \ldots, \mathcal{K}_{n_j}^j\}$, $j \in \mathbb{N}$ denote the family of triangulation where j is the refinement step. We say that the mesh size of the triangulation $\mathcal{T}$ at the jth refinement step is h^j if

$$h^j = \max_{\mathcal{K}_i^j \in \mathcal{T}^j} h_{\mathcal{K}_i^j}, \quad \text{and} \quad h_{\mathcal{K}_i^j} = \operatorname{diam}(\mathcal{K}_i^j).$$

Here $\operatorname{diam}(\mathcal{K}_i^j) = \sup_{\mathbf{x}, \mathbf{y} \in \mathcal{K}_i^j} \|\mathbf{x} - \mathbf{y}\|_2$.

The quantity h^j is a measure of how refined the mesh is. The smaller h^j is, the finer the mesh. Let $\rho_{\mathcal{K}_i}$ denote the radius of inscribed circle in $\mathcal{K}_i$. We say that a family of triangulation $\mathcal{T}^j$, $j \in \mathbb{N}$ is *shape regular* if there exists a constant $\kappa > 0$, independent of j such that for each $\mathcal{K}_i^j \in \mathcal{T}^j$, we have [74]

$$\frac{h_{\mathcal{K}_i^j}}{\rho_{\mathcal{K}_i}^j} \leq \kappa.$$

For this problem, we generate a shape regular mesh using distmesh [70].

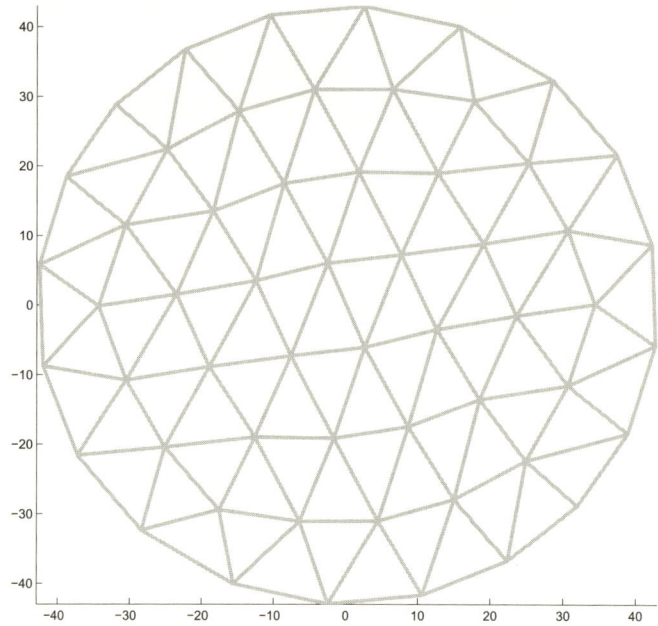

Fig. 14.3. The shape regular mesh.

Usually one chooses the finite dimensional subspace V, to be the space of continuous piecewise polynomials on the domain Ω_h. Piecewise linear or quadratic functions are widely used in defining these subspaces. Let $\{B_i\}_{i=1}^{N_u}$ be a basis for the finite dimensional space V, then the discrete weak formulation is expressed as a linear system of equations in the following form:

$$AC_f = R. \tag{14.13}$$

Here the coefficients of the matrix A are given by $A_{ij} = \mathcal{D}_{ij} + \mathcal{M}_{ij} + \mathcal{P}_{ij}$, where the diffusion inner product $\mathcal{D}_{ij} = \int_{\Omega_h} \nabla B_i \cdot D \nabla B_j \, d\varsigma \, dv$, the absorption inner product $\mathcal{M}_{ij} = \int_{\Omega_h} B_i \mu_a B_j \, d\varsigma \, dv$, and the boundary inner product $\mathcal{P}_{ij} = \frac{1}{2} \int_{\partial \Omega_h} B_i B_j \, ds$. The components of the forcing vector R_i in the right-hand side are $\int_{\Omega_h} B_i f \, d\varsigma \, dv$, and C_f is the N_u-vector containing the unknown coefficients $c_i^f : C_f = (c_1^f, c_2^f, \ldots, c_{N_u}^f)^T$.

14.4 The inverse problem

To obtain simulation data for the inverse problem for optical tomography, we solve the forward problem (14.4), (14.5) using a high accuracy finite element Galerkin method and known spatial distributions for D and μ_a [76]. For a set of source functions f_j this yields *exact* measurement values of u on the boundaries, denoted

14.4 The inverse problem

by $g_{\delta_{ij}}$. In the inverse algorithm, the forward problem (14.4), 14.5 is always solved using fewer basis polynomials for u, D and μ_a, namely smaller N_u, N_D, and N_μ. For a given distribution for D and μ_a, this yields a set of observables $\mathcal{C}_{ij}$ which approximate $g_{\delta_{ij}}$ and depend (*nonlinearly*) on the coefficients $\mathbf{q} \in \mathbb{R}^{N_t}$, where $N_t = N_\mu + N_D$.

The goal of the inverse problem is then to obtain the distribution for D and μ_a given measurements of u on the boundaries for applied sources f_j. This can be cast as a finite dimensional minimization problem

$$\frac{1}{2}\min_{\mathbf{q}\in\mathbb{R}^N} J(\mathbf{q}) = \frac{1}{2}\min_{\mathbf{q}} \sum_{j=1}^{n_s}\sum_{i=1}^{n_m} (\mathcal{C}_{ij}(u(\mathbf{x}_i;\mathbf{q},f_j)) - z_{\delta_{ij}})^2$$

$$= \frac{1}{2}\min_{\mathbf{q}} \sum_{p=1}^{n_t} (\mathcal{C}_p(u(\mathbf{q})) - z_{\delta_p})^2$$

$$= \frac{1}{2}\min_{\mathbf{q}} \|F(\mathbf{q})\|^2. \qquad (14.14)$$

Here the residual vector $F(\mathbf{q}) \in \mathbb{R}^{n_t}$, $n_t = n_s \times n_m$, comprises all terms $\mathcal{C}_{ij} - z_{\delta_{ij}}$. Problem (14.14) can be solved by any iteratively regularized method discussed in the previous sections. At every step, given values for the coefficients of basis-function expansions, $\mathbf{q}_n$, the unknown spatial distributions for D and μ_a may be reconstructed through those expansions, i.e.,

$$D(\mathbf{x}) = \sum_{l=1}^{N_D} (Q_D)_l B_l(\mathbf{x}), \quad \mu_a(\mathbf{x}) = \sum_{l=1}^{N_\mu} (Q_\mu)_l B_l(\mathbf{x}), \qquad (14.15)$$

where $\mathbf{q}_n = (Q_D, Q_\mu)^T = (\mathbf{q}^{(1)}, \mathbf{q}^{(2)}, \ldots, \mathbf{q}^{(N_t)})^T$. A typical observation operator is

$$\mathcal{C}_{ij}(u(\mathbf{q})) = -D\frac{\partial u}{\partial v}(\mathbf{x}_i;\mathbf{q},f_j) = \frac{1}{2}u(\mathbf{x}_i;\mathbf{q},f_j), \qquad (14.16)$$

where $i = 1, 2, \ldots, n_m$, $j = 1, 2, \ldots, n_s$, $\mathbf{x}_i$ is in $\partial\Omega$, n_m is the number of measurements, n_s is the number of sources, and the second equality comes from the boundary condition (14.5).

We further suppose that finite dimensional problem (14.14) adequately describes the modeled process, and our goal becomes to solve problem (14.14) specifically. In order to implement algorithms (10.3), (10.38) and (9.5) one has to compute $F'(\mathbf{q}) = K(\mathbf{q})$, the Jacobian matrix of $F(\mathbf{q})$. For observation operator (14.16), one gets [14]

$$K(\mathbf{q}) = \left\{-D\frac{\partial}{\partial q^{(k)}}\left(\frac{\partial u_p}{\partial v}(\mathbf{q})\right)\right\}_{p=\overline{1,n_t},\,k=\overline{1,N_t}}$$

$$= \left\{\frac{1}{2}\frac{\partial u_p}{\partial q^{(k)}}(\mathbf{q})\right\}_{p=\overline{1,n_t},\,k=\overline{1,N_t}} \qquad (14.17)$$

14 Inverse problem in optical tomography

with $u_{i+n_m(j-1)} = u(\mathbf{x}_i; \mathbf{q}, f_j)$. If we plug the expression

$$u_{i+n_m(j-1)} = \sum_{l=1}^{N_u} c_l^{f_j} B_l(\mathbf{x}_i)$$

into equation (14.17) then

$$[K(\mathbf{q})]_{pk} = \frac{1}{2} \sum_{l=1}^{N_u} \frac{\partial c_l^{f_j}}{\partial \mathbf{q}^{(k)}} B_l(\mathbf{x}_i) = \mathbf{B}^T \frac{\partial C_{f_j}}{\partial \mathbf{q}^{(k)}}, \quad p = i + n_m(j-1), \qquad (14.18)$$

where

$$\mathbf{B} = \frac{1}{2}(B_1(\mathbf{x}_i), B_2(\mathbf{x}_i), \ldots, B_{N_u}(\mathbf{x}_i)). \qquad (14.19)$$

Substitute expressions (14.15) for $D(\mathbf{x})$ and $\mu_a(\mathbf{x})$ into the diffusion

$$\mathcal{D}_{ij} = \int_\Omega \nabla B_i \cdot D \nabla B_j \, d\varsigma \, dv$$

and the absorption

$$\mathcal{M}_{ij} = \int_\Omega B_i \mu_a B_j \, d\varsigma \, dv$$

inner products, respectively. One obtains

$$\mathcal{D}_{ij} = \sum_{k=1}^{N_D} (Q_D)_k \mathbb{D}_{k,ij}, \quad \text{where } \mathbb{D}_{k,ij} = \int_\Omega B_k \nabla B_i \cdot \nabla B_j \, d\varsigma \, dv, \qquad (14.20)$$

and similarly

$$\mathcal{M}_{ij} = \sum_{k=1}^{N_\mu} (Q_\mu)_k \mathbb{M}_{k,ij}, \quad \text{where } \mathbb{M}_{k,ij} = \int_\Omega B_k B_i B_j \, d\varsigma \, dv. \qquad (14.21)$$

Now differentiate equation (14.13) with respect to the parameters $\mathbf{q}^{(k)}$ assuming that R is independent of $\mathbf{q}^{(k)}$. We get

$$A \frac{\partial C_f}{\partial \mathbf{q}^{(k)}} + \frac{\partial A}{\partial \mathbf{q}^{(k)}} C_f = 0. \qquad (14.22)$$

Solve for the partial derivatives with respect to the parameters to obtain for $N_t = N_D + N_\mu$

$$\frac{\partial C_f}{\partial \mathbf{q}^{(k)}} = -(\mathcal{D}(\mathbf{q}) + \mathcal{M}(\mathbf{q}) + \mathcal{P})^{-1} \mathbb{D}_k C_f(\mathbf{q}), \quad 1 \le k \le N_D, \qquad (14.23)$$

$$\frac{\partial C_f}{\partial \mathbf{q}^{(k)}} = -(\mathcal{D}(\mathbf{q}) + \mathcal{M}(\mathbf{q}) + \mathcal{P})^{-1} \mathbb{M}_k C_f(\mathbf{q}), \quad N_D + 1 \le k \le N_t, \qquad (14.24)$$

where $A = \mathcal{D} + \mathcal{M} + \mathcal{P}$, $\mathbb{D}_k$ and $\mathbb{M}_k$ are the $N_u \times N_u$ matrices with elements $\mathbb{D}_{k,ij}$ and $\mathbb{M}_{k,ij}$, respectively. Thus using equations (14.23) and (14.24) for the expression $\partial C_{f_j}/\partial \mathbf{q}^{(k)}$ in equation (14.18) we can derive the Jacobian matrix $K(\mathbf{q})$ for our inverse calculation.

The explicit expression for $K(\mathbf{q})$ yields $\nabla J = K^*(\mathbf{q})F(\mathbf{q})$, and the approximation of the first term of the Hessian with $\nabla^2 J(\mathbf{q}) \approx K^*(\mathbf{q})K(\mathbf{q})$. Then, iteratively regularized Gauss–Newton (IRGN) method (10.3) for solving (14.14) is given by

$$\mathbf{q}_{n+1} = \mathbf{q}_n + \mathbf{p}_n, \qquad (14.25)$$

where the search direction $\mathbf{p}_n$ is a solution to the linear system

$$(K^*(\mathbf{q}_n)K(\mathbf{q}_n) + \alpha_n I)\mathbf{p}_n = -(K^*(\mathbf{q}_n)F(\mathbf{q}_n) + \alpha_n(\mathbf{q}_n - \xi)), \qquad (14.26)$$

and I is the identity matrix. To implement (10.38), the following inner iterations are to be performed for every $n = 0, 1, \ldots, n(\delta)$

$y_0 = 0$
for $m = 0$ to $M(n) - 1$ do
$\quad y_{m+1} = (I - \vartheta K^*(\mathbf{q}_n)K(\mathbf{q}_n))y_m + \vartheta K^*(\mathbf{q}_n)\{F(\mathbf{q}_n) - K(\mathbf{q}_n)(\mathbf{q}_n - \xi)\}.$
endfor

Then $\mathbf{q}_{n+1} = \xi - y_{M(n)}$.

For $1/\alpha_n = M(n)$ and ϑ any real number from the interval $(0, 2/N_1^2)$, $\|K(\mathbf{q}_n))\|_2 \leq N_1$, this corresponds to

$$\Theta(F'^*(\mathbf{q}_n)F'(\mathbf{q}_n), \alpha_n) = \vartheta \sum_{k=0}^{M(n)} (I - \vartheta F'^*(\mathbf{q}_n)F'(\mathbf{q}_n))^k$$

in (10.38). In what follows we will refer to this method as iteratively regularized Gauss–Landweber (IRGL). In our numerical experiment, the efficiency of the above two methods will be compared to the efficiency of iteratively regularized gradient (IRG) scheme (9.5). Being applied to (14.14) with $\gamma_n = \vartheta \alpha_n$, it takes the form

$$\mathbf{q}_{n+1} = \mathbf{q}_n - \vartheta \alpha_n \{K^*(\mathbf{q}_n)F(\mathbf{q}_n) + \alpha_n(\mathbf{q}_n - \xi)\}. \qquad (14.27)$$

14.5 Numerical results

In solving both the forward and the inverse problems, we assume our domain $\Omega \subset \mathbb{R}^2$ to be a disc of radius 43.0 mm centered at the origin. In our simulations, we use a 32 equidistant source-detector configuration similar to the experimental setup in [50]. In this setup, one uses 16 equidistant sources ($n_s = 16$) at the boundary and 16 alternate equidistant detectors ($n_m = 16$) at the boundary, where the measurements

120   14 Inverse problem in optical tomography

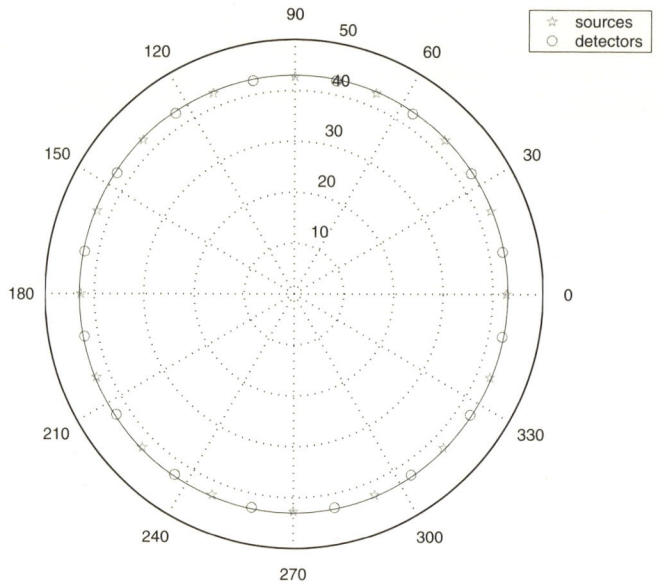

Fig. 14.4. The source-detector configuration.

are taken, see Figure 14.4. Therefore, our inverse problem solver uses $16 \times 16 = 256$ simulated data for the reconstruction of the diffusion and absorption coefficients.

For the comparison of different iterative methods, we simulate a circular inclusion of radius 10 mm located at the point $\mathbf{x}_0 = (15, 15)$. For the forward problem, the absorption coefficient μ_a is taken to be 0.012 mm^{-1} inside the inclusion and 0.006 mm^{-1} outside as a background value. The diffusion coefficient D is taken to be 0.275 mm^{-1} inside and 0.55 mm^{-1} as a background value, see Figure 14.5. In our simulations, the sources are modeled as a narrow Gaussian, mainly $f(\mathbf{x}) = (A/\sqrt{2\pi\varepsilon^2})\exp(-\|\mathbf{x} - \mathbf{x}_s\|_2^2/2\varepsilon^2)$ for a source located at $\mathbf{x}_s$, to mimic laser sources where typical values of $A = 1.0 \times 10^5$ and $\varepsilon = 0.1$ are used as in [50].

Inversion results with $N_u = 128$ basis functions for u, and 128 basis functions for each of D and μ_a are given, although the algorithm is also successful for $N_D \neq N_\mu$ and for $N_D \neq N_u$. Each iteration is initiated with the diffusion and absorption spatial distributions at the background, non-tumor values. Simulation with known parameters for D and μ_a leads to a calculable residual of $\delta \approx 10^{-5}$. For (10.3) and (10.38) we take $\alpha_n = 2^{-n}$. Note, that by stopping rule

$$\alpha_n = \delta, \tag{14.28}$$

the total number of Newton steps for both methods (10.3) and (10.38) is $n = n(\delta) = 17$.

For iteratively regularized gradient method (9.5), $n(\delta) = \frac{1}{\delta^2} = 10^{10}$, since α_n has to decrease "slowly" and we take $\alpha_n = \frac{1}{\sqrt{n}}$. The approximate values of the dif-

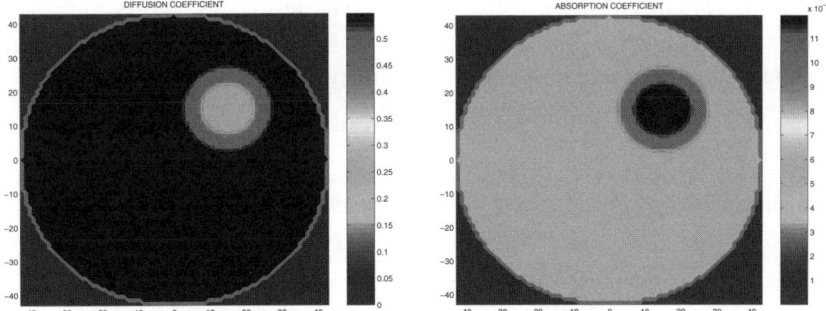

Fig. 14.5. The diffusion and absorption coefficients. Exact values.

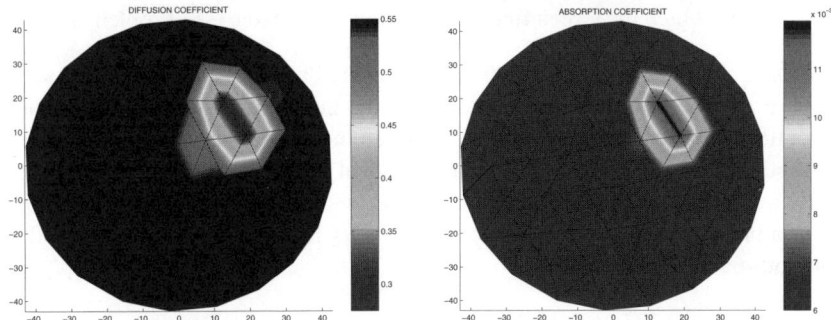

Fig. 14.6. The diffusion and absorption coefficients. Computed values.

fusion and absorption coefficients obtained by iteratively regularized Gauss–Newton method (10.3) are illustrated in Figure 14.6.

In conclusion, we would like to point out that this is just a preliminary experiment, which illustrates how iteratively regularized methods work for parameter identification problems in PDEs. A lot more research is required in order to improve the quality of numerical reconstructions and to make the jump from experimental systems to a fully accepted clinical tool.

Problems

14.1. Calculate two analytical solutions to problem (14.8)–(14.9), which can be used to test the forward solver in 1D case (see Problem 14.3). Namely, solve problem (14.8)–(14.9) with $f(\mathbf{x}) = e^{-x}$ and $f(\mathbf{x}) = \delta(\mathbf{x} - \mathbf{x}_s)$. Take $L = 43.0$ mm, $\mathbf{q} = \sqrt{\mu_a/D}$, $\mu_a = 0.012$ mm^{-1}, and $D = 0.33$ mm.

14.2. Using integration by parts, write 1D time-independent problem (14.8)–(14.9) in form (14.12).

14.3. Write a computer code to solve 1D forward problem (14.8)–(14.9) numerically by finite element method described in Section 14.3. To test your forward solver, first compute approximate solutions in the following two cases:

1. constant values of D and μ_a ($D = 0.33$ mm and $\mu_a = 0.012$ mm^{-1}) and $f(\mathbf{x}) = e^{-x}$, $\mathbf{x} \in (0, 43.0)$,

2. constant values of D and μ_a ($D = 0.33$ mm and $\mu_a = 0.012$ mm^{-1}) and

$$f(\mathbf{x}) = \frac{A}{\sqrt{2\pi\epsilon^2}} \exp\left(-\frac{|\mathbf{x} - \mathbf{x}_s|^2}{2\epsilon^2}\right), \quad \mathbf{x} \in (0, 43.0), \qquad (14.29)$$

with $A = 1$ and $\epsilon = 0.1$, where $x_s = 42.5$ is the source location. This particular f is an approximated delta distributed source (DDS), i.e., approximation to $f(\mathbf{x}) = \delta(\mathbf{x} - \mathbf{x}_s)$, for which the Green function solution is available (see Problem 14.1).

14.4. Once the forward solver has been tested (see Problems 14.1-14.3), it is safe to compute z_{ij} in order to provide the data for the inverse problem. It is well known that optical contrast is mainly determined by an increase in haemoglobin concentration around cancerous tissue which increases local absorption μ_a and decreases local scattering D. Taking into consideration possible size, position, and optical properties of a breast tumor, chose contrast values of μ_a and D between background and cancerous tissues as follows:

$$D(x) := \begin{cases} 0.275, & x \in [1, 10], \quad \text{tumor } D(x) \text{ value} \\ 0.550, & x \in [0, 1) \text{ and } x \in (10, 43], \quad \text{background } D(x) \text{ value} \end{cases}$$

$$\mu_a(x) := \begin{cases} 0.012, & x \in [1, 10], \quad \text{tumor } \mu_a(x) \text{ value} \\ 0.006, & x \in [0, 1) \text{ and } x \in (10, 43], \quad \text{background } \mu_a(x) \text{ value}. \end{cases}$$

Use the above parameters together with $f_j(\mathbf{x}) = \frac{A}{\sqrt{2\pi\epsilon^2}} \exp\left(-\frac{|\mathbf{x} - \mathbf{x}_s^j|^2}{2\epsilon^2}\right)$, $A = 1$, $\epsilon = 0.1$, to generate the values of z_{ij} in (14.7). Here $\mathbf{x}_s^j$ are the source locations. For your numerical simulations, use 20 different sources ($j = 1, 2, \ldots, 20$) located close to the boundary. Namely,

$$0 < \mathbf{x}_s^1 < \mathbf{x}_s^2 < \cdots < \mathbf{x}_s^{10} \leq 0.5 \quad \text{and} \quad 42.5 \leq \mathbf{x}_s^{11} < \mathbf{x}_s^{12} < \cdots < \mathbf{x}_s^{20} < 43.0.$$

14.5. Apply iteratively regularized procedures (10.3), (10.38) and (9.5) to reconstruct the profiles for μ_a and D in one-dimensional case (see Section 14.4 for details).

15

Feigenbaum's universality equation

15.1 The universal constants

In this chapter we investigate *Feigenbaum's universality equation* [32]

$$g_z(x) = \alpha_z g_z\left(g_z\left(\frac{x}{\alpha_z}\right)\right) \tag{15.1}$$

associated with the iteration theory of real functions. A solution g_z to this nonlinear equation describes the local structure of attractors. It also defines two *universal constants* α_z and δ_z. For the iterative process

$$x_{k+1} = f_{\mu,z}(x_k) \tag{15.2}$$

with a differentiable transition function $f_{\mu,z}$, which has a unique maximum $\hat{x}$ such that

$$f_{\mu,z}(x) - f_{\mu,z}(\hat{x}) \approx |x - \hat{x}|^z, \tag{15.3}$$

the universal constants α_z and δ_z have the following meaning [32]:

$$\alpha_z = \lim_{j \to \infty} \frac{d_j}{d_{j+1}}, \tag{15.4}$$

$$\delta_z = \lim_{j \to \infty} \frac{\mu_j - \mu_{j-1}}{\mu_{j+1} - \mu_j}. \tag{15.5}$$

Here d_j is the algebraic distance between $\hat{x}$ and the nearest attractor, the limit point in 2^j cycle. The value of d_j is negative if the limit point is less than $\hat{x}$. In the second identity, μ_j is the critical value of μ, for which the doubling of the period (the appearance of 2^j cycle) of the transition function $f_{\mu,z}$ occurs.

It is important to mention that while the two constants α_z and δ_z depend on the value of z in (15.3), they do not depend on the function $f_{\mu,z}$ itself. In that sense, α_z

15 Feigenbaum's universality equation

and δ_z are universal. Consider, for example, the following iterative schemes [78] on the interval $[0, 1]$:

$$x_{k+1} = \mu x_k (1 - x_k), \tag{15.6}$$

and

$$x_{k+1} = \mu \sin(\pi x_k). \tag{15.7}$$

In the first case, the transition function $f_{\mu,2}(x) := \mu x(1 - x)$. In the second case, $f_{\mu,2}(x) := \mu \sin(\pi x)$. The dynamical systems generated by these iterations, change from periodic to chaotic through the doubling of a period as μ takes certain critical values $\mu_1, \mu_2, \ldots, \mu_j, \ldots$. Specifically, one can verify that for $f_{\mu,2}(x) := \mu x(1-x)$ and $\mu = \mu_1 = 3$, the fixed points ξ_1 and ξ_2 of $f_{\mu,2}$ become unstable and two more points ξ_3 and ξ_4, such that

$$f_{\mu,2}(\xi_3) = \xi_4, \quad \text{and} \quad f_{\mu,2}(\xi_4) = \xi_3,$$

appear. This is called 2^1 cycle. Then for $\mu = \mu_2 = 3.449499\ldots$, the points ξ_3 and ξ_4 become unstable and the points ξ_5, ξ_6, ξ_7 and ξ_8, such that

$$f_{\mu,2}(\xi_5) = \xi_6, \quad f_{\mu,2}(\xi_6) = \xi_7, \quad f_{\mu,2}(\xi_7) = \xi_8, \quad \text{and} \quad f_{\mu,2}(\xi_8) = \xi_5,$$

appear. This is called 2^2 cycle. In general, for $\mu = \mu_j$ the points of 2^{j-1} cycle become unstable and 2^j cycle appears. Remarkably, the values of α_2 and δ_2, which characterize this process, are the same for both transition functions. The first seven critical values of μ (with six decimal places) [78] for (15.6) are

$$\mu_2 = 3.449499, \quad \mu_3 = 3.544090, \quad \mu_4 = 3.564407, \quad \mu_5 = 3.568759,$$
$$\mu_6 = 3.569692, \quad \mu_7 = 3.569891, \quad \mu_8 = 3.569934.$$

Notice that

$$\lim_{j \to \infty} \mu_j := \mu_\infty = 3.569946.$$

When $\mu > \mu_\infty$, the behavior of system (15.6) becomes chaotic. Further research shows [78] that two-dimensional Hannon's scheme

$$\begin{cases} x_{k+1} = 1 - \mu x_k^2 + y_k \\ y_{k+1} = b x_k \end{cases} \tag{15.8}$$

also exhibits the above bifurcation phenomena (for some fixed values of b) and δ_2 is the same as for (15.6) and (15.7).

15.2 Ill-posedness

Due to extreme sensitivity of computed solutions to the rounding errors, Feigenbaum's equation is ill-posed (unstable). In addition to that, uniqueness of the solution can only be guaranteed on a certain class of analytic functions. One can easily verify that if $g_z(x)$ is a solution to (15.1) and γ is a constant, then any $f_z(x) = \gamma g_z(\frac{x}{\gamma})$ is a solution as well [78]. Indeed,

$$\alpha_z f_z\left(f_z\left(\frac{x}{\alpha_z}\right)\right) = \alpha_z \gamma g_z\left(\frac{f_z(\frac{x}{\alpha_z})}{\gamma}\right) = \alpha_z \gamma g_z\left(\frac{\gamma g_z(\frac{x}{\alpha_z \gamma})}{\gamma}\right)$$

$$= \gamma\left[\alpha_z g_z\left(g_z\left(\frac{x}{\alpha_z \gamma}\right)\right)\right] = \gamma g_z\left(\frac{x}{\gamma}\right) = f_z(x). \quad (15.9)$$

Therefore one has to normalize the solution by setting, for example, $g_z(0) = 1$ [22]. This implies

$$\alpha_z = \frac{1}{g_z(1)}. \quad (15.10)$$

Combining (15.1) and (15.10), one arrives at the following equation

$$F(g_z) := g_z(1) g_z(x) - g_z(g_z(g_z(1)x)) = 0. \quad (15.11)$$

It has been shown in [22] that a solution to (15.11) is still not unique. Each solution has power series expansion in the form

$$g_z(x) = 1 + \sum_{i=1}^{\infty} q_i |x|^{zi} \quad (15.12)$$

for some $1 < z < \infty$. The corresponding universal constant $\alpha_z = \frac{1}{g_z(1)}$ reflects the structure of attractors in the dynamical system with nonlinearity of order z, see [78]. The computer-assisted proof of the existence result can be found in [55].

15.3 Numerical algorithm for $2 \leq z \leq 12$

Let us discuss a numerical method for solving equation (15.11) and computing the first universal constant α_z. The values of α_z are practically important, but still very little is known about them. For instance, it is not clear if they satisfy any simple algebraic relations.

To obtain the data for verifying different hypotheses concerning α_z, K. Briggs [22] calculated α_z for integer z, $2 \leq z \leq 12$, with high accuracy on a class of even concave down functions, analytic on $[-1, 1]$. In [22], approximate solutions to (15.11) are constructed as polynomials:

$$g_z(x) = 1 + \sum_{i=1}^{n} q_i |x|^{zi}, \quad (15.13)$$

where the coefficients q_i are obtained from the following nonlinear system:

$$F_j := \left(1 + \sum_{i=1}^{n} q_i\right)\left(1 + \sum_{i=1}^{n} q_i |x_j|^{zi}\right) - 1$$

$$-\sum_{i=1}^{n} q_i \left|1 + \sum_{i=1}^{n} q_i\left|(1 + \sum_{i=1}^{n} q_i)x_j\right|^{zi}\right|^{zi} = 0, \quad j = 1, \ldots, n. \quad (15.14)$$

Since g_z is even, it only has to be evaluated on $[0, 1]$. For $z = 2$ and $z = 3$, the choice of grid points is not very important from computational point of view. Therefore the uniform partition $x_j = \frac{j}{n}$, $j = 1, 2, \ldots, n$, is used. However for $z \geq 4$, the nonlinear partition $x_j = \left(\frac{j}{n}\right)^{1/z}$ gives more accurate results. The Jacobian matrix

$$F' := \left[\frac{\partial F_j}{\partial q_l}\right]_{j,l=\overline{1,n}}$$

$$= |x_j|^{zl}\left(1 + \sum_{i=1}^{n} q_i\right)$$

$$+ \left(1 + \sum_{i=1}^{n} q_i |x_j|^{zi}\right) - \left|1 + \sum_{i=1}^{n} q_i\left|(1 + \sum_{i=1}^{n} q_i)x_j\right|^{zi}\right|^{zl}$$

$$- \text{sign}(h_j) \sum_{i=1}^{n} q_i z i \left|1 + \sum_{i=1}^{n} q_i\left|(1 + \sum_{i=1}^{n} q_i)x_j\right|^{zi}\right|^{zi-1} \quad (15.15)$$

$$\times \left[\left|(1 + \sum_{i=1}^{n} q_i)x_j\right|^{zl} + \text{sign}(g_z(1))\sum_{i=1}^{n} q_i z i \left|(1 + \sum_{i=1}^{n} q_i)x_j\right|^{zi-1}|x_j|\right]$$

is strictly ill-posed for $z \geq 2$ and $n \geq 2$. Here

$$h_j := 1 + \sum_{i=1}^{n} q_i \left|(1 + \sum_{i=1}^{n} q_i)x_j\right|^{zi}.$$

The condition number for any fixed n increases about ten times as z is replaced with $z + 1$. K. Briggs [22] conducted the experiment with 200 significant digits in order to reduce the rounding error and overcome the instability of system (15.14). As the result, he was able to successfully calculate α_z for $2 \leq z \leq 12$ by classical Newton's method. Below are the values of α_z, $2 \leq z \leq 12$, with five decimal places [22]

$\alpha_2 = -2.50290, \quad \alpha_3 = -1.92769, \quad \alpha_4 = -1.69030, \quad \alpha_5 = -1.55577,$
$\alpha_6 = -1.46774, \quad \alpha_7 = -1.40511, \quad \alpha_8 = -1.35802, \quad \alpha_9 = -1.32119,$
$\alpha_{10} = -1.29152, \quad \alpha_{11} = -1.26706, \quad \alpha_{12} = -1.24653.$

The solution to system (15.14) for $z = 2$ and $n = 7$ with nine decimal places is as follows [78]:

$$g_2(x) := 1 - 1.527632997x^2 + 0.104815194x^4 + 0.026705673x^6$$
$$- 0.003527413x^8 + 0.000081581x^{10}$$
$$+ 0.000025368x^{12} - 0.000002687x^{14}. \tag{15.16}$$

The corresponding value of $\alpha_2 = \frac{1}{g_2(1)}$ is $-2.502907875\ldots$. For $z \geq 13$, despite the high accuracy of computations, Newton's method diverges. The system becomes unstable to the point that standard numerical algorithms are no longer applicable.

15.4 Regularized method for $z \geq 13$

The goal of the numerical experiment performed by the authors was, first, to solve nonlinear system (15.14) for $2 \leq z \leq 12$ and to get the results consistent with [22] and, second, to calculate α_z for $z \geq 13$. The authors would like to thank Dr. V. Vasin for introducing them to the above problem and for helping the authors to carry the numerical simulations out.

To stabilize universality equation (15.11), the iteratively regularized Gauss–Newton (IRGN) algorithm, presented in Chapter 10, was implemented:

$$x_{k+1} = \xi - (F'^*(x_k)F'(x_k) + \lambda_k E_1)^{-1} F'^*(x_k)[F(x_k) - F'(x_k)(x_k - \xi)]$$

with $\lambda_k = \frac{\lambda_0}{k}$. For large values of z, the initial regularization parameter λ_0 increases by a factor of 10 at every transition from z to $z + 1$ to prevent the Jacobian from becoming singular to working precision. All the experiments presented in this section were conducted with 24 significant digits.

For a more accurate approximation of g_z one has to take n large enough, but then the problem of the choice of an initial approximation occurs: for $z = 2$, $n = 2$ or $n = 3$ system (15.14) has many solutions. For this reason the scheme described in [9] is used. First, system (15.14) is solved with $n = 1$, then the solution to (15.14), $n = 1$, is taken as the initial guess for the case $n = 2$. The solution with $n = 2$ is taken as the initial guess for $n = 3$, etc [1]. When $z = 2$, $n = 1$, $x = 1$ system (15.14) is reduced to one algebraic equation with respect to q_1:

$$q_1(q_1 + 1)(q_1^5 + 3q_1^4 + 3q_1^3 + 3q_1^2 + 2q_1 - 1) = 0$$

and the two obvious solutions are $q_1^{(1)} = 0$, $q_1^{(2)} = -1$. Since the function $g_z(x)$ is even and concave down, the initial condition $g_z(0) = 1$ implies $g_z(1) < 1$, that is $1 + q_1 < 1$, $q_1 < 0$. Therefore one has to find the negative roots of the equation:

$$q_1^5 + 3q_1^4 + 3q_1^3 + 3q_1^2 + 2q_1 - 1 = 0.$$

Such roots are: $q_1^{(3)} = -1.8597174\ldots$, $q_1^{(4)} = -1.4021968\ldots$. Thus for the system of two equations ($z = 2$, $n = 2$) the initial data are:

1. $q_1 = -1, q_2 = 0$;
2. $q_1 = -1.8597174\ldots, q_2 = 0$;
3. $q_1 = -1.4021968\ldots, q_2 = 0$.

In the first two cases the solutions to (15.14) ($z = 2, n = 2$) are not concave down on $[-1, 1]$. In the third case the graph of the polynomial is concave down and

$$g_2(x) \approx 1 - 1.5416948x^2 + 0.1439197x^4.$$

For the system of three equations ($z = 2, n = 3$) the initial data are:

$$q_1 = -1.5416948, \quad q_2 = 0.1439197, \quad q_3 = 0.$$

Then we continue this process. The maximum dimension we take is $n = 12$. If $n = 13$, the discrepancy is not less than for $n = 12$, and after $n = 14$ it grows.

For $z = 3$ we begin the computation with one equation ($n = 1$) also. As the initial approximation $q_1 = -1.4021968$ is taken, that is the solution to (15.14) with $z = 2, n = 1$. The dimension increases step by step till the discrepancy improves. For $z = 3, n = 1$ the solution to (15.14) with $z = 3, n = 1$ is used, etc. In our experiment α_z for $z = 2, \ldots, 26$ are found. For $z = 2, \ldots, 12$ they coincide with α_z given in [22]. Below the values of $\alpha_{13}, \alpha_{14}, \ldots, \alpha_{26}$ are presented.

$\alpha_{13} = -1.22902, \quad \alpha_{14} = -1.21391, \quad \alpha_{15} = -1.20072, \quad \alpha_{16} = -1.18910,$
$\alpha_{17} = -1.17879, \quad \alpha_{18} = -1.16957, \quad \alpha_{19} = -1.1612, \quad \alpha_{20} = -1.1537,$
$\alpha_{21} = -1.1469, \quad \alpha_{22} = -1.140, \quad \alpha_{23} = -1.134, \quad \alpha_{24} = -1.129,$
$\alpha_{25} = -1.124, \quad \alpha_{26} = -1.12.$

The above experiment can be considered successful. Due to regularization and step-by-step choice of the initial guess, ill-conditioned nonlinear system (15.14) has been solved for $2 \leq z \leq 26$. The universal constants α_z have been calculated.

Problems

15.1. Apply *iteratively regularized Newton's method*:

$$x_{k+1} = x_k - [F'(x_k) + \lambda_k E_1]^{-1}(F(x_k) + \lambda_k(x_k - \xi)), \quad k = 1, 2, \ldots$$

to solve nonlinear system (15.14) numerically. Use step-by-step algorithm [9], described in Chapter 15, to find initial approximations for $z = 2, 3, \ldots$. If possible, take $\lambda_k = 0$ for small values of z.

15.2. Check that the Jacobian matrix is given by formula (15.15).

15.3. In [32], it has been shown that the second universal constant δ_z is the largest modulus eigenvalue of the operator $L(g_z) : C^2(-\infty, \infty) \to C(-\infty, \infty)$, where

$$L(g_z) = T'(g_z), \quad T(g_z) := \alpha_z g_z \left(g_z \left(\frac{x}{\alpha_z} \right) \right).$$

Prove that [22, 78]

$$L(g_z)h = \alpha_z \left\{ g'_z \left(g_z \left(\frac{x}{\alpha_z} \right) \right) \cdot h \left(\frac{x}{\alpha_z} \right) + h \left(g_z \left(\frac{x}{\alpha_z} \right) \right) \right\}.$$

15.4. By Problem 15.3, the universal constant δ_z is a solution to the eigenvalue problem

$$L(g_z)h = \delta_z h.$$

As suggested in [78], one can calculate an approximate value of δ_z by setting $h(x) = $ const. Verify that in this case

$$\alpha_z(g'_z(1) + 1) = \delta_z.$$

15.5. Show that

$$g'_z(x) = g'_z \left(g_z \left(\frac{x}{\alpha_z} \right) \right) \cdot g'_z \left(\frac{x}{\alpha_z} \right),$$

$$g''_z(x) = g''_z \left(g_z \left(\frac{x}{\alpha_z} \right) \right) \cdot \left(g'_z \left(\frac{x}{\alpha_z} \right) \right)^2 \frac{1}{\alpha_z} + g'_z \left(g_z \left(\frac{x}{\alpha_z} \right) \right) \cdot g''_z \left(\frac{x}{\alpha_z} \right) \frac{1}{\alpha_z}.$$

Derive from the last equation that

$$\alpha_z = g'_z(1) \quad \text{and} \quad \alpha_z^2 + \alpha_z = \delta_z.$$

16

Conclusion

Let us make some methodological remarks on the material presented. Throughout the course, in addition to studying irregular equations $F(x) = 0$, $x \in H_1$, with exact operators $F : H_1 \to H_2$, we investigate the equations

$$F(x) = f, \quad x \in H_1, \tag{16.1}$$

which contain some noise in the right-hand side. Namely, in place of the element $f \in H_2$, a function $\tilde{f}$ is given such that

$$\|\tilde{f} - f\|_{H_2} \leq \delta. \tag{16.2}$$

The value of δ is also assumed to be known. In this case, it is not possible to solve equation (16.1) exactly. One can only try to find some approximations that are close in a certain sense to the solution set

$$X^*(F) = \{x \in H_1 : F(x) = f\},$$

and converge to this set as $\delta \to 0$ regardless of the choice of $\tilde{f}$ from condition (16.2). Convergence of the approximations to $X^*(F)$ is usually understood as convergence to zero of the distance $\mathrm{dist}(\,\cdot\,, X^*(F))$ along the sequence of these approximations. While constructing the required approximations, one has to use all a priori information available about the exact right-hand side f, i.e., the function $\tilde{f}$ and the error estimate δ. Formally, any such method can be associated with a mapping $\mathcal{R}$, which takes a pair of input data $(\tilde{f}, \delta)$, $\|\tilde{f} - f\|_{H_2} \leq \delta$, to some approximate solution $x(\tilde{f}, \delta) = \mathcal{R}(\tilde{f}, \delta)$ of equation (16.1). The convergence assumption can now be written in the form

$$\lim_{\delta \to 0} \sup_{\tilde{f} : \|\tilde{f} - f\|_{H_2} \leq \delta} \mathrm{dist}(x(\tilde{f}, \delta), X^*(F)) = 0. \tag{16.3}$$

A mapping $\mathcal{R} : (\tilde{f}, \delta) \to x(\tilde{f}, \delta)$, satisfying property (16.3), is called a *regularizing algorithm* for original equation (16.1). If a regularizing algorithm $\mathcal{R}$ has been constructed for this equation and, instead of the exact element f, the corresponding pair $(\tilde{f}, \delta)$ is known, then the point $\mathcal{R}(\tilde{f}, \delta)$ can be taken as an approximate solution to (16.1).

16 Conclusion

In Chapters 5 and 6 there are examples of regularizing algorithms for irregular equations with affine operators F (see (5.26), (6.20)). In a nonlinear case, a regularizing algorithm, suitable for effective numerical implementation, can be produced if the operator F generates a convex discrepancy functional (Problem 9.2)

$$\Phi(x) = \frac{1}{2}\|F(x) - f\|_{H_2}^2.$$

As for equations with arbitrary irregular nonlinear operators $F \in \mathcal{F}(N_1, N_2)$, presently, Tikhonov scheme (4.18) is the only source of regularizing algorithms, theoretically appropriate for solving such equations. The fact that the mapping $\mathcal{R}(\tilde{f}, \delta) = x_{\alpha(\delta)}^{\varepsilon(\delta)}$, obtained by this scheme, is a regularizing algorithm follows from (4.22). However, as it is pointed out in Chapter 4, there are principal difficulties in practical realization of Tikhonov's scheme, since the Tikhonov functional Φ_α to be minimized may have numerous local minima.

Evidently, equations with affine operators and with the operators leading to convex discrepancy functionals exhaust the list of equations that can be effectively treated by Tikhonov's scheme. Therefore, one has to admit that the above definition of a regularizing algorithm is too restrictive in a general nonlinear case in a sense that it is hard to find at least one algorithm satisfying this definition, which can be performed numerically for any $F \in \mathcal{F}(N_1, N_2)$.

In Chapters 10 and 11 an alternative approach to the construction of regularizing algorithms is presented. This approach uses the idea of iterative methods equipped with special control parameters operating in such a way that identity (16.3) holds for some acceptable values of these parameters only. In our methods, the element ξ works as a control parameter. The ellipsoid, defined by condition (10.6) (or (10.24)), see Chapter 10, or the finite dimensional affine subspace M_{x^*}, see Chapter 11, play a part of the admissible set Ξ. Here it is convenient to assume that $X^*(F) = \{x^*\}$. It follows from estimates (10.34) and (11.34), where one has to put $\Delta = 0$, that iterations (10.30) and (11.32) generate regularizing algorithms if $\xi \in \Xi$.

Since in both our cases Ξ depends on the unknown solution x^* and its interior is empty, the problem of finding $\xi \in \Xi$ in practice is hardly simpler than the original equation. Thus, normally, instead of $\xi \in \Xi$, only some element ξ, located more or less close to $\Xi = \Xi(x^*)$, can be available, and one has to figure out how much this affects the quality of the approximations. Estimates (10.34) and (11.34) show that the error, introduced for the above reason, is a function of the distance $\Delta = \text{dist}(\xi, \Xi)$ and converges to zero as $\Delta \to 0$. In other words, the iterative methods, considered in Chapters 10 and 11, are stable not only with respect to errors in the right-hand side f, but also with respect to deviations of the element ξ from the admissible set Ξ.

Finally, note that the problem of finding ξ from a small neighborhood of the set Ξ is very similar to the one of choosing an initial approximation from a small neighborhood of the solution for locally convergent iterative processes. As we have seen in Chapters 1–3, problems of this kind occur even in case of iterative methods applied to regular operator equations.

References

[1] Airapetyan, R. G., Ramm, A. G., Smirnova, A. B.: Continuous methods for solving nonlinear ill-posed problems, in: Operator theory and applications, pp. 111–138, Amer. Math. Soc., Fields Institute Communications, Providence, RI (2000).

[2] Akimova, E. N., Vasin V. V.: Stable parallel algorithms for solving the inverse gravimetry and magnetometery problems, CD Proceedings of the 9th International Conference on Numerical Methods in Continuum Mechanics, Zilina, Slovakia, September 9–12 (2003).

[3] Alber, Ya., Nashed, Z.: Iterative-projection regularization of ill-posed variational inequalities. Analysis (Munich) **24**:1 (2004), 19–39.

[4] Alber, Ya., Ryazantseva, I.: Nonlinear Ill-Posed Problems of Monotone Type. Springer, Dordrecht (2006).

[5] Allaire, G., Kaber, S. M.: Numerical Linear Algebra. Texts in Applied Mathematics, Springer, New York (2008).

[6] Andrieux, S., Baranger, T. N., Ben Abda, A.: Solving Cauchy problems by minimizing an energy-like functional. Inverse Problems **22**:1 (2006), 115–133.

[7] Arridge, S. R.: Optical tomography in medical imaging: Topical review, Inverse Problems **15** (1999), 41–93.

[8] Atkinson, K. E.: An Introduction to Numerical Analysis, John Wiley and Sons, New York (1978).

[9] Babenko, K. I.: Fundamentals of Numerical Analysis, Nauka, Moscow (1986).

[10] Bakushinsky, A. B., Goncharsky, A. V.: Ill-Posed Problems: Theory and Applications, Kluwer, Dordrecht (1994).

[11] Bakushinsky, A. B. Iterative methods without saturation for solving degenerate nonlinear operator equations, Dokl. Russian Acad. Sci. **334** (1995), 7–8.

[12] Bakushinsky, A. B.: Iterative methods with fuzzy feedback for solving irregular operator equations, Dokl. Russian Acad. Sci. **428**:5 (2009), 1–3.

[13] Bakushinsky, A. B., Kokurin, M. Yu.: Iterative methods for Ill-Posed Operator Equations with Smooth Operators, Springer Dordrecht, Great Britain (2004).

[14] Bakushinsky, A. B., Khan, T., Smirnova, A.: Inverse Problem in Optical Tomography and its Numerical Investigation by Iteratively Regularized Methods, Journal of Inverse and Ill-posed Problems **13**:4 (2005), 1–14.
[15] Bakushinsky, A. B., Smirnova, A., Skinner, M. A.: Iteratively regularized gradient method with a posteriori stopping rule for 2D inverse gravimetry problem, Journal of Integral Equations and Applications **17** (2005), 375–390.
[16] Bakushinsky, A. B., Smirnova, A.: On application of generalized discrepancy principle to iterative methods for nonlinear ill-posed problems, Numerical Functional Analysis and Optimization **26** (2005), 35–48.
[17] Bakushinsky, A. B., Smirnova, A., Tuncer, N.: Relative computational efficiency of iteratively regularized methods. Journal of Inverse and Ill-posed Problems **16**:4 (2008), 1–14.
[18] Bauer, F., Hohage, T.: A Lepskij-type stopping rule for regularized Newton methods, Inverse Problems **21** (2005), 1975–1991.
[19] Bauer, F., Hohage, T., Munk, A.: Iteratively regularized Gauss–Newton method for nonlinear inverse problems with random noise. SIAM J. Numer. Anal. **47**:3 (2009), 1827–1846.
[20] Braess, D.: Finite Elements. Theory, fast solvers, and applications in solid mechanics. Cambridge (2001).
[21] Brenner, S., Ridgway, S. L.: The mathematical theory of finite element methods, Springer (1994).
[22] Briggs, K.: A precise calculation of the Feigenbaum constants, Mathematics of computations **57**:195 (1991), 435–439.
[23] Burger, M., Osher, S. J.: A survey on level set methods for inverse problems and optimal design. European J. Appl. Math. **16**:2 (2005), 263–301.
[24] Chance B., Alfano, R. R.: Optical tomography, photon migration, and spectroscopy of tissue and model media: theory, human studies, and instrumentations, part 1 and 2, SPIE **2389** (1995).
[25] Chapko, R., Johansson, B.: Tomas An alternating boundary integral based method for a Cauchy problem for the Laplace equation in semi-infinite regions. Inverse Probl. Imaging **2**:3 (2008), 317–333.
[26] Ciarlet., P. E.: The Finite Element Method for Elliptic Problems. SIAM (2002).
[27] Colton, D., Kress, R.: Inverse Acoustic and Electromagnetic Scattering Theory. Second edition. Applied Mathematical Sciences, 93. Springer, Berlin (1998).
[28] Deimling, K.: Nonlinear functional analysis, Springer, New York (1985).
[29] Dennis, J. E. Schnabel, R. B. Numerical Methods for Unconstrained Optimization and Nonlinear Equations, Prentice-Hall, Englewood Cliffs, NJ (1983).
[30] Egger, H., Engl, H. W.: Tikhonov regularization applied to the inverse problem of option pricing: convergence analysis and rates. Inverse Problems **21**:3 (2005), 1027–1045.
[31] Engl, H., Hanke, M., Neubauer, A.: Regularization of Inverse Problems, Kluwer Academic Publisher, Dordecht, Boston, London (1996).
[32] Feigenbaum, M. J.: Quantitative of universality for a class of nonlinear transformations, J. Stat. Phys. **19**:1 (1978), 25–52.

[33] Gongsheng, L., Nashed, Z.: A modified Tikhonov regularization for linear operator equations. Numer. Funct. Anal. Optim. **26**:4–5 (2005), 543–563.

[34] Griesbaum, A., Kaltenbacher, B., Vexler, B.: Efficient computation of the Tikhonov regularization parameter by goal-oriented adaptive discretization. (English summary) Inverse Problems **24**:2 (2008), 20.

[35] Groetsch, C. W.: Inverse problems in mathematical sciences, Vieweg, Braunschweig (1993).

[36] Groetsch, C. W.: Regularization and stabilization of inverse problems. Handbook of analytic-computational methods in applied mathematics, 31–64, Chapman and Hall/CRC, Boca Raton, FL (2000).

[37] Hanke, M., Nagy, J., Vogel, C.: Quasi-Newton approach to nonnegative image restorations, Linear Algebra and Its Applications **316** (2000), 223–236.

[38] Hielscher, A., Alcouffe, R., Barbour, R.: Comparison of finite-difference transport and diffusion calculations for photon migration in homogenous and heterogenous tissues, Phys. Med. Biol. **42** (1998), 1285–1302.

[39] Hohage, T: Logarithmic convergence rates of the iteratively regularized Gauss–Newton method for an inverse potential and inverse scattering problem, Inverse problems **13** (1997), 1279–1299.

[40] Isakov, V.: Inverse problems for Partial Differential Equations. Second edition. Applied Mathematical Sciences, 127. Springer, New York (2006).

[41] Jin, Q.: A conergence analysis of the iteratively regularized Gauss–Newton method under the Lipschitz condition, Inverse problems **24**:4 (2008), 16pp.

[42] Kabanikhin, S. I., Lorenzi, A.: Identification problems of wave phenomena. Theory and numerics. Inverse and Ill-posed Problems Series. VSP, Utrecht (1999).

[43] Kabanikhin, S. I., Satybaev, A. D., Shishlenin, M. A.: Direct Methods of Solving Multidimensional Inverse Hyperbolic Problems. Inverse and Ill-posed Problems Series. VSP, Utrecht (2005).

[44] Kaltenbacher, B.: A note on logarithmic convergence rates for nonlinear Tikhonov regularization. J. Inverse Ill-Posed Probl. **16**:1 (2008), 79–88.

[45] Kaltenbacher, B., Hofmann, B.: Convergence rates for the iteratively regularized Gauss–Newton method in Banach spaces. Inverse Problems **26**:3, 21 (2010).

[46] Kaltenbacher, B., Neubauer, A., Scherzer, O.: Iterative Regularization Methods for Nonlinear Ill-Posed Problems, Radon Series on Computational and Applied Mathematics, de Gruyter, Berlin (2008).

[47] Kaltenbacher, B., Schöpfer, F., Schuster, T.: Iterative methods for nonlinear ill-posed problems in Banach spaces: convergence and applications to parameter identification problems. Inverse Problems **25**:6 (2009), 065003.

[48] Kantorovich, L. V., Akilov, G. P.: Functional Analysis in Normed Spaces, MacMillan Co., New York (1964).

[49] Khan, T., Smirnova, A.: 1D inverse problem in diffusion based optical tomography using iteratively regularized Gauss–Newton algorithm, pp. 149–170, Applied Mathematics and Computation, Inc., 161. Elsevier Science (2005).

[50] Khan, T., Thomas, A.: Inverse problem in refractive index based optical tomography. Inverse Problems **22**:4 (2006), 1121–1137.

[51] Kirsch, A.: An introduction to the mathematical theory of inverse problems, Springer, New York (1996).

[52] Kokurin, M. Yu.: Stable iteratively regularized gradient method for nonlinear irregular equations under large noise. Inverse Problems **22**:1 (2006), 197–207.

[53] Kozlov, V. A., Maz′ya, V. G.: Iterative procedures for solving ill-posed boundary value problems that preserve the differential equations. (Russian) Algebra i Analiz **1**:5 (1989), 144–170; translation in Leningrad Math. J. **1**:5 (1990), 1207–1228.

[54] Kozlov, V. A., Maz′ya, V. G., Fomin, A. V.: An iterative method for solving the Cauchy problem for elliptic equations. (Russian) Zh. Vychisl. Mat. i Mat. Fiz. **31**:1, 64–74 (1991); translation in U.S.S.R. Comput. Math. and Math. Phys. **31**:1 (1991), 45–52.

[55] Lanford, O. E.: A computer-assisted proof of the Feigenbaum conjectures, Bull. Amer. Math. Soc. **6**:3 (1982), 427–434.

[56] Langemann, D., Tasche, M.: Phase reconstruction by a multilevel iteratively regularized Gauss–Newton method. Inverse Problems **24**:3 (2008), 26.

[57] Langer, S.: Complexity analysis of the iteratively regularized Gauss–Newton method with inner CG-iteration. J. Inverse Ill-Posed Probl. **17**:9 (2009), 671–890.

[58] Langer, S., Hohage, T.: Convergence analysis of an inexact iteratively regularized Gauss–Newton method under general source conditions. J. Inverse Ill-Posed Probl. **15**:3 (2007), 311–327.

[59] Lattès, R., Lions, J.-L.: The Method of Quasi-Reversibility. Applications to Partial Differential Equations. Translated from the French edition and edited by Richard Bellman. Modern Analytic and Computational Methods in Science and Mathematics, 18. American Elsevier Publishing Co., Inc., New York (1969).

[60] Mahale, P., Nair, M.: Thamban A simplified generalized Gauss–Newton method for nonlinear ill-posed problems. Math. Comp. **78**:256 (2009), 171–184.

[61] Marin, L., An iterative MFS algorithm for the Cauchy problem associated with the Laplace equation. CMES Comput. Model. Eng. Sci. **48**:2 (2009), 121–153.

[62] Markel, V. A., Schotland, J. C.: On the convergence of the Born series in optical tomography with diffuse light. Inverse Problems **23** N4 (2007), 1445–1465.

[63] Melnikova, I. V., Filinkov, A.: Abstract Cauchy Problems: Three Approaches. Chapman & Hall/CRC Monographs and Surveys in Pure and Applied Mathematics, 120. Chapman & Hall/CRC, Boca Raton, FL (2001).

[64] Nash, S. G., Sofer, A.: Linear and Nonlinear Programming, McGraw-Hill, New York (1996).

[65] Natterer, F., Wubbeling.F.: Mathematical Methods in Image Reconstruction, SIAM Monographs on Mathematical Modelling and Computation (2001).

[66] Nicaise, S.: About the Lamé system in a polygonal or a polyhedral domain and a coupled problem between the Lamé system and the plate equation. I. Regu-

larity of the solutions. Ann. Scuola Norm. Sup. Pisa Cl. Sci. (4) **19**:3 (1992), 327–361.

[67] Nocedal, J., Wright, S. J.: Numerical Optimization, Springer, New York (1999).

[68] Ortega, J. M., Rheinboldt, W. C.: Iterative Solution of Nonlinear Equations in Several Variables, Academic Press (1970).

[69] Osher, S., Fedkiw, R.: Level set Methods and Dynamic Implicit Surfaces. Applied Mathematical Sciences, 153. Springer, New York (2003).

[70] Perrson, P., Strang, G.: A simple mesh generator in MATLAB. SIAM Review **46**:2 (2004), 329–345.

[71] Schechter, M.: An introduction to nonlinear analysis. Cambridge University Press, Cambridge (2004).

[72] Scherzer, O.: A Modified Landweber Iteration for Solving Parameter Estimation Problems, Appl. Math.& Optim. **38** (1998), 45–68.

[73] Schotland, J. C.: Inversion formulas and fast image reconstruction algorithms in optical tomography. Mathematical methods in biomedical imaging and intensity-modulated radiation therapy (IMRT), 367–387, CRM Series, 7, Ed. Norm., Pisa (2008).

[74] Shewchuk, J. R.: What is a good linear element? Interpolation, conditioning, and quality measures, Eleventh international meshing roundtable (Ithaca, New York), 115–126, Sandia National Laboratories (2002).

[75] Smirnova, A. B., Renaut R. A.: A family of preconditioned iteratively regularized methods for nonlinear minimization. Journal of Inverse and Ill-Posed Problems. **17**:4 (2009), 405–418.

[76] Smirnova, A. B., Renaut R. A., Khan, T.: Convergence and application of a modified iteratively regularized Gauss–Newton algorithm. Inverse Problems **23**:4 (2007), 1547–1563.

[77] Vasin, V. V., Ageev, A. L.: Ill-posed problems with a priori information, VNU, Utrecht (1995).

[78] Vasin, V. V., Ryashko, L. B.: Elements of Nonlinear Dynamics: from Order to Chaos (in Russian), Ural State University, Ekaterinburg (2003).

[79] Vogel, C. R., Numerical solution of a nonlinear ill-posed problem arizing in inverse scattering. Inverse Problems **1** (1985), 393–403.

[80] Vogel, C. R.: Computational Methods for Inverse Problems. Frontiers in Applied Mathematics. Society for Industrial and Applied Mathematics (SIAM), Philadelphia (2002).

[81] Yagola, A., Titarenko, V.: Using a priori information about a solution of an ill-posed problem for constructing regularizing algorithms and their applications. Inverse Probl. Sci. Eng. **15**:1 (2007), 1–17.

Index

compact embedding condition   27

diffusion approximation   111
discrepancy principle   42

equation
   Feigenbaum's universality   123
   irregular   3
   radiative transfer   111
   regular   3, 11

Fréchet derivative   1
functional
   strictly convex   21
   strongly convex   22, 32
   Tikhonov   23

generalized discrepancy principle   104

ill-posed problem   3
iterative regularization   70

least squares solution   10
linearization   2

method
   finite element   114, 122
   gradient   16
   iteratively regularized gradient   70
   Landweber   16
   Newton's   4
   steepest descent   16
   the Gauss–Newton   12
   the generalized Gauss–Newton   84
   the iteratively regularized Gauss–Newton   77
   the iteratively regularized Newton's   128
minimizing sequence   24

operator
   compact   9
   continuously invertible   3
   regular   3, 11
   strongly continuous   25
   weakly closed   30
   weakly continuous   25

problem
   inverse gravimetry   103, 109
   inverse magnetometery   110
   inverse scattering   109
   nonlinear least square   112

quasisolution   10

regularizing algorithm   130
reverse connection   84

saturation phenomena   61
source-type condition   56, 77

universal constants   123

weakly compact set   25